U.S.NRC
United States Nuclear Regulatory Commission
Protecting People and the Environment

NUREG-1437, Volume 3
Revision 1

Generic Environmental Impact Statement for License Renewal of Nuclear Plants

Appendices

Final Report

Office of Nuclear Reactor Regulation

AVAILABILITY OF REFERENCE MATERIALS IN NRC PUBLICATIONS

NRC Reference Material

As of November 1999, you may electronically access NUREG-series publications and other NRC records at NRC's Public Electronic Reading Room at http://www.nrc.gov/reading-rm.html. Publicly released records include, to name a few, NUREG-series publications; *Federal Register* notices; applicant, licensee, and vendor documents and correspondence; NRC correspondence and internal memoranda; bulletins and information notices; inspection and investigative reports; licensee event reports; and Commission papers and their attachments.

NRC publications in the NUREG series, NRC regulations, and Title 10, "Energy," in the *Code of Federal Regulations* may also be purchased from one of these two sources.

1. The Superintendent of Documents
   U.S. Government Printing Office
   Mail Stop SSOP
   Washington, DC 20402–0001
   Internet: bookstore.gpo.gov
   Telephone: 202-512-1800
   Fax: 202-512-2250
2. The National Technical Information Service
   Springfield, VA 22161–0002
   www.ntis.gov
   1–800–553–6847 or, locally, 703–605–6000

A single copy of each NRC draft report for comment is available free, to the extent of supply, upon written request as follows:
Address: U.S. Nuclear Regulatory Commission
Office of Administration
Publications Branch
Washington, DC 20555-0001
E-mail: DISTRIBUTION.RESOURCE@NRC.GOV
Facsimile: 301–415–2289

Some publications in the NUREG series that are posted at NRC's Web site address http://www.nrc.gov/reading-rm/doc-collections/nuregs are updated periodically and may differ from the last printed version. Although references to material found on a Web site bear the date the material was accessed, the material available on the date cited may subsequently be removed from the site.

Non-NRC Reference Material

Documents available from public and special technical libraries include all open literature items, such as books, journal articles, transactions, *Federal Register* notices, Federal and State legislation, and congressional reports. Such documents as theses, dissertations, foreign reports and translations, and non-NRC conference proceedings may be purchased from their sponsoring organization.

Copies of industry codes and standards used in a substantive manner in the NRC regulatory process are maintained at—
The NRC Technical Library
Two White Flint North
11545 Rockville Pike
Rockville, MD 20852–2738

These standards are available in the library for reference use by the public. Codes and standards are usually copyrighted and may be purchased from the originating organization or, if they are American National Standards, from—
American National Standards Institute
11 West 42nd Street
New York, NY 10036–8002
www.ansi.org
212–642–4900

Legally binding regulatory requirements are stated only in laws; NRC regulations; licenses, including technical specifications; or orders, not in NUREG-series publications. The views expressed in contractor-prepared publications in this series are not necessarily those of the NRC.

The NUREG series comprises (1) technical and administrative reports and books prepared by the staff (NUREG–XXXX) or agency contractors (NUREG/CR–XXXX), (2) proceedings of conferences (NUREG/CP–XXXX), (3) reports resulting from international agreements (NUREG/IA–XXXX), (4) brochures (NUREG/BR–XXXX), and (5) compilations of legal decisions and orders of the Commission and Atomic and Safety Licensing Boards and of Directors' decisions under Section 2.206 of NRC's regulations (NUREG–0750).

United States Nuclear Regulatory Commission

Protecting People and the Environment

NUREG-1437, Volume 3
Revision 1

Generic Environmental Impact Statement for License Renewal of Nuclear Plants

Appendices

Final Report

Manuscript Completed: May 2013
Date Published: June 2013

Office of Nuclear Reactor Regulation

Cover Sheet

Responsible Agency: U.S. Nuclear Regulatory Commission, Office of Nuclear Reactor Regulation

Title: *Final Generic Environmental Impact Statement for License Renewal of Nuclear Plants* (NUREG-1437) Volumes 1, 2, and 3, Revision 1

For additional information or copies of this Final Generic Environmental Impact Statement for License Renewal of Nuclear Plants, contact:

Division of License Renewal
U.S. Nuclear Regulatory Commission
Office of Nuclear Reactor Regulation
Mail Stop O-11F1
11555 Rockville Pike
Rockville, Maryland 20852
Phone: 1-800-368-5642, extension 1183
Fax: (301) 415-2002
Email: LRGEISUpdate@nrc.gov

Abstract

U.S. Nuclear Regulatory Commission (NRC) regulations allow for the renewal of commercial nuclear power plant operating licenses. To support the license renewal environmental review process, the NRC published the *Generic Environmental Impact Statement for License Renewal of Nuclear Plants* (GEIS) in 1996. The proposed action considered in the GEIS is the renewal of nuclear power plant operating licenses.

Since publication of the GEIS, approximately 40 plant sites (70 reactor units) have applied for license renewal and undergone environmental reviews, the results of which were published as supplements to the 1996 GEIS. This GEIS revision reviews and reevaluates the issues and findings of the 1996 GEIS. Lessons learned and knowledge gained during previous license renewal reviews provide a significant source of new information for this assessment. In addition, new research, findings, public comments, and other information were considered in evaluating the significance of impacts associated with license renewal.

The intent of the GEIS is to determine which issues would result in the same impact at all nuclear power plants and which issues could result in different levels of impact at different plants and thus require a plant-specific analysis for impact determinations. The GEIS revision identifies 78 environmental impact issues for consideration in license renewal environmental reviews, 59 of which have been determined to be generic to all plant sites. The GEIS also evaluates a full range of alternatives to the proposed action. For most impact areas, the proposed action would have impacts that would be similar to or less than impacts of the alternatives, in large part because most alternatives would require new power plant construction, whereas the proposed action would not.

Paperwork Reduction Act Statement
This NUREG contains information collection requirements that are subject to the Paperwork Reduction Act of 1995 (44 USC 3501 et seq.). These information collections were approved by the Office of Management and Budget, approval numbers 3150-0021.

Public Protection Notification
The NRC may not conduct or sponsor, and a person is not required to respond to, a request for information or an information collection requirement unless the requesting document displays a currently valid OMB control number.

Contents

Figures

Figures

Tables

Tables

Tables

Tables

Acronyms, Abbreviations, and Chemical Nomenclature

ABWR	advanced boiling water reactor
AC	alternating current
ACRS	Advisory Committee on Reactor Safeguards
ACS	American Cancer Society
ADAMS	Agencywide Documents Access and Management System
AEA	Atomic Energy Act
AEC	U.S. Atomic Energy Commission
AGNIR	Advisory Group on Non-ionizing Radiation
AIRFA	American Indian Religious Freedom Act of 1978
ALARA	as low as is reasonably achievable
ALI	annual limit on intake
APE	Area of Potential Effect
ALWR	advanced light water reactor
ASLB	Atomic Safety and Licensing Board
ASME	American Society of Mechanical Engineers
AWEA	American Wind Energy Association
BEIR	Biological Effects of Ionizing Radiation (National Research Council Committee)
BLM	U.S. Bureau of Land Management
BLS	U.S. Bureau of Labor Statistics
BMPs	best management practices
BPA	Bonneville Power Administration
BWR	boiling water reactor
BOEMRE	Bureau of Ocean Energy Management, Regulation, and Enforcement
CAA	Clean Air Act
CADHS	California Department of Health Services
CCS	carbon capture and storage
CCW	coal combustion waste
CDC	Centers for Disease Control and Prevention
CDF	core damage frequency
CdTe	cadmium telluride
CEC	California Energy Commission
CEDE	committed effective dose equivalent

CEG	Constellation Energy Group
CEQ	Council on Environmental Quality
CERCLA	Comprehensive Environmental Response, Compensation, and Liability Act
CF	capacity factor
CFR	*Code of Federal Regulations*
CGEC	California Geothermal Energy Collaborative
CH_4	methane
CHP	combined heat and power
CIGS	copper-indium-gallium-selenide
CLB	current licensing basis
CO	carbon monoxide
CO_2	carbon dioxide
COL	combined operating license
CSP	concentrating solar power
CWA	Clean Water Act
CZMA	Coastal Zone Management Act
DC	direct current
DDREF	dose and dose rate effectiveness factor
DNC	Dominion Nuclear Connecticut
DNI	direct normal insolation
DOD	U.S. Department of Defense
DOE	U.S. Department of Energy
DOL	U.S. Department of Labor
DSM	demand-side management
EA	environmental assessment
EAB	exclusion area boundary
ECRR	European Committee on Radiation Risk
EEI	Edison Electric Institute
EERE	Energy Efficiency and Renewable Energy
EEZ	Exclusive Economic Zone
EF	enhanced Fujita (scale)
EFH	essential fish habitat
EGS	engineered geothermal systems
EI	exposure index
EIA	Energy Information Administration
EIML	Environmental Incorporated Midwest Laboratory
EIS	environmental impact statement
EJ	environmental justice
ELF-EMF	extremely low frequency-electromagnetic field

EMF	electromagnetic field
EMF-RAPID	Electric and Magnetic Fields Research and Public Information Dissemination (Program)
EP	emergency planning
EPA	U.S. Environmental Protection Agency
EPAct	Energy Policy Act of 2005
EPCRA	Emergency Planning and Community Right-to-Know Act
EPRI	Electric Power Research Institute
ER	environmental report
ERCOT	Electric Reliability Council of Texas
ERO	Electric Reliability Organization
ESA	Endangered Species Act
ESP	early site permit
Exelon	Exelon Generating Company LLC
F	Fujita (scale)
FAA	Federal Aviation Administration
FCC	Federal Communications Commission
FDOH	Florida Department of Health
FEMA	Federal Emergency Management Agency
FERC	Federal Energy Regulatory Commission
FES	final environmental statement
FGD	flue gas desulfurization
FICN	Federal Interagency Committee on Noise
FIFRA	Federal Insecticide, Fungicide, and Rodenticide Act
FPL	Florida Power & Light Company
FR	*Federal Register*
FRCC	Florida Reliability Coordinating Council
FSAR	Final Safety Analysis Report
FS	U.S. Forest Service
GALL	Generic Aging Lessons Learned
GAO	U.S. General Accounting Office (now U.S. Government Accountability Office)
GCRP	U.S. Global Change Research Program
GDC	General Design Criterion
GEA	Geothermal Energy Association
GEIS	generic environmental impact statement
GHG	greenhouse gas
GIS	geographic information system
GNEP	Global Nuclear Energy Partnership

GSMFC	Gulf States Marine Fisheries Commission
GTCC	greater than Class C
HAP	hazardous air pollutant
HAPC	habitat area of particular concern
HAWT	horizontal axis wind turbine
HCCP	Harvard Center for Cancer Prevention
HDR	hot dry rock
HFC	hydrofluorocarbon
HCFC	hydrochlorofluorocarbon
HHV	higher heating value
HLW	high-level (radioactive) waste
HVAC	heating, ventilation, and air conditioning
IAEA	International Atomic Energy Agency
IARC	International Agency for Research on Cancer
ICM	Interim Compensatory Measure
ICRP	International Commission on Radiological Protection
IDPH	Illinois Department of Public Health
IDNR	Idaho Department of Natural Resources
IEEE	Institute of Electrical and Electronic Engineers
IGCC	integrated gasification combined cycle
IMP	Indiana Michigan Power
INIRC	International Non-Ionizing Radiation Commission
IPE	Individual Plant Examination
IPEEE	Individual Plant Examination of External Events
IRPA	International Radiation Protection Association
ISFSI	independent spent fuel storage installation
ISI	in-service inspection
IWSA	Integrated Waste Services Association
LERF	large early release frequency
LET	linear energy transfer
LFG	landfill gas
LLAP	*Legionella*-like amoebal pathogen
LLD	lower limit of detection
LLNL	Lawrence Livermore National Laboratory
LLW	low-level (radioactive) waste
LLRWPA	Low-Level Radioactive Waste Policy Act
LLTF	Lessons Learned Task Force
LLWPAA	Low-Level Radioactive Waste Policy Act Amendments

LOA	letter of authorization
LOEL	lowest observed effects level
LWR	light water reactor
MACCS	MELCOR Accident Consequence Code System
MACT	maximum achievable control technology
MCAQD	Maricopa County Air Quality Department
MCL	maximum contaminant level
MEI	maximally exposed individual
MMPA	Marine Mammal Protection Act
MMS	Minerals Management Service
MSFCMA	Magnuson-Stevens Fishery Conservation and Management Act
MSW	municipal solid waste
MTBE	methyl tertiary butyl ether
NAAQS	National Ambient Air Quality Standards
NaCl	sodium chloride (salt)
NAICS	North American Industry Classification System
NAGPRA	Native American Graves Protection and Repatriation Act
$NaNO_3$	sodium nitrate
NAS	National Academy of Sciences
National Register	*National Register of Historic Places*
NCDC	National Climatic Data Center
NCRP	National Council on Radiation Protection and Measurements
NEI	Nuclear Energy Institute
NEPA	National Environmental Policy Act of 1969
NERC	North American Electric Reliability Corporation
NESC	National Electrical Safety Code
NETL	National Energy Technology Laboratory
NGCC	natural gas combined cycle
NGL	natural gas liquids
NHPA	National Historic Preservation Act of 1966
$(NH_4)SO_4$	ammonium sulfate
NIEHS	National Institute of Environmental Health Sciences
NIH	National Institutes of Health
NJDEP	New Jersey Department of Environmental Protection
NMC	Nuclear Management Company
NMFS	National Marine Fisheries Service
NO	nitrogen oxide
N_2O	nitrous oxide
NO_2	nitrogen dioxide

NOAA	National Oceanic and Atmospheric Administration
NORM	naturally occurring radioactive material
NOS	National Oceanic Service
NO_x	nitrogen oxides
NPCC	Northeast Power Coordinating Council
NPDES	National Pollutant Discharge Elimination System
NPP	nuclear power plant
NRC	U.S. Nuclear Regulatory Commission
NREL	National Renewable Energy Laboratory
NRPB	National Radiological Protection Board
NSPS	New Source Performance Standards
NWI	National Waste Initiative; National Wetland Inventory
NWPA	National Waste Policy Act
NYSDEC	New York State Department of Environmental Conservation
NYSDOL	New York State Department of Labor
O_3	ozone
OCS	Outer Continental Shelf
ODCM	Offsite Dose Calculation Manual
OPPD	Omaha Public Power District
OTA	Office of Technology Assessment
OSHA	Occupational Safety and Health Administration
PAH	polycyclic aromatic hydrocarbon
PARS	Publicly Available Record System
Pb	lead
PC	pulverized coal
PCB	polychlorinated biphenyl
PDR	Public Document Room
PEIS	programmatic environmental impact statement
PFC	perfluorocarbon
PI	performance indicator
PILOT	payments in lieu of tax
PM	particulate matter
$PM_{2.5}$	particulate matter with a mean aerodynamic diameter of 2.5 μm or less
PM_{10}	particulate matter with a mean aerodynamic diameter of 10 μm or less
POTW	publicly owned treatment works
PPE	personal protective equipment
PRA	probabilistic risk assessment
PSD	prevention of significant deterioration
PTC	production tax credit

PURPA	Public Utility Regulatory Act of 1978
PV	photovoltaic
PVC	photovoltaic cell
PWR	pressurized water reactor
RCRA	Resource Conservation and Recovery Act of 1976
RD&D	research, development, and demonstration
RDF	refuse-derived fuel
REMP	Radiological Environmental Monitoring Program
RER	radiological effluent release
RERR	radiological effluent release report
RES	Renewable Energy Standard
RFC	Reliability First Corporation
ROP	Reactor Oversight Program
ROW	right-of-way
RPS	Renewable Portfolio Standards
RNA	ribonucleic acid
RRC	Regional Reliability Council
RRY	reference reactor year
SAAQS	State Ambient Air Quality Standards
SAMA	severe accident mitigation alternative
SAMDA	severe accident mitigation design alternative
SCE	Southern California Edison
SCR	selective catalytic reduction
SDWA	Safe Drinking Water Act
SEIS	supplemental environmental impact statement
SER	safety evaluation report
SFP	spent fuel pool
SHPO	State Historic Preservation Office or Officer
SIP	State implementation plan
SMITTR	surveillance, monitoring, inspection, testing, trending, and recordkeeping
SNYPSC	State of New York Public Service Commission
SO_2	sulfur dioxide
SOARCA	state-of-the-art reactor consequence analysis
SPAR	standardized plant analysis risk
SPDES	State Pollutant Discharge Elimination System
SPP	Southwest Power Pool
SRM	Staff Requirements Memorandum
SSCs	systems, structures, and components

Stat.	*Statutes at Large*
STG	steam turbine generator
TCEQ	Texas Commission on Environmental Quality
TDS	total dissolved solids
TEDE	total effective dose equivalent
TESS	threatened and endangered species system
THPO	Tribal Historic Preservation Officer
TLD	thermoluminescence dosimeter
TSCA	Toxic Substances Control Act
TSS	total suspended solids
TTU	Texas Tech University
TVA	Tennessee Valley Authority
TXU	TXU Generation Company
UCB	upper confidence bound
UCS	Union of Concerned Scientists
UF_6	uranium hexafluoride
UNSCEAR	United Nations Scientific Committee on the Effects of Atomic Radiation
UO_2	uranium dioxide
U_3O_8	triuranium octaoxide
USACE	U.S. Army Corps of Engineers
USC	*United States Code*
USCB	U.S. Census Bureau
USDA	U.S. Department of Agriculture
USFWS	U.S. Fish and Wildlife Service
USGS	U.S. Geological Survey
VOC	volatile organic compound
WCNOC	Wolf Creek Nuclear Operating Corporation
WCS	Waste Control Specialists LLC
WEC	wave energy capture
WGA	Western Governors' Association
WHO	World Health Organization

Shortened Nuclear Power Plant Names Used in This Report

Arkansas	Arkansas Nuclear One
Beaver Valley	Beaver Valley Power Station
Braidwood	Braidwood Station
Browns Ferry	Browns Ferry Nuclear Plant
Brunswick	Brunswick Steam Electric Plant
Byron	Byron Station
Callaway	Callaway Plant
Calvert Cliffs	Calvert Cliffs Nuclear Power Plant
Catawba	Catawba Nuclear Station
Clinton	Clinton Power Station
Columbia	Columbia Generating Station
Comanche Peak	Comanche Peak Steam Electric Station
Cooper	Cooper Nuclear Station
Crystal River	Crystal River Nuclear Power Plant
Cook	Donald C. Cook Nuclear Plant
Davis-Besse	Davis-Besse Nuclear Power Station
Diablo Canyon	Diablo Canyon Power Plant
Dresden	Dresden Nuclear Power Station
Arnold	Duane Arnold Energy Center
Farley	Joseph M. Farley Nuclear Plant
Fermi	Enrico Fermi Atomic Power Plant
FitzPatrick	James A. FitzPatrick Nuclear Power Plant
Fort Calhoun	Fort Calhoun Station
Ginna	R.E. Ginna Nuclear Power Plant
Grand Gulf	Grand Gulf Nuclear Station
Harris	Shearon Harris Nuclear Power Plant
Hatch	Edwin I. Hatch Nuclear Plant
Hope Creek	Hope Creek Generating Station
Indian Point	Indian Point Energy Center
Kewaunee	Kewaunee Power Station
LaSalle	LaSalle County Station
Limerick	Limerick Generating Station
McGuire	McGuire Nuclear Station
Millstone	Millstone Power Station
Monticello	Monticello Nuclear Generating Plant
Nine Mile Point	Nine Mile Point Nuclear Station
North Anna	North Anna Power Station
Oconee	Oconee Nuclear Station
Oyster Creek	Oyster Creek Nuclear Generating Station

Palisades	Palisades Nuclear Plant
Palo Verde	Palo Verde Nuclear Generating Station
Peach Bottom	Peach Bottom Atomic Power Station
Perry	Perry Nuclear Power Plant
Pilgrim	Pilgrim Nuclear Power Station
Point Beach	Point Beach Nuclear Plant
Prairie Island	Prairie Island Nuclear Generating Plant
Quad Cities	Quad Cities Nuclear Power Station
River Bend	River Bend Station
Robinson	H.B. Robinson Steam Electric Plant
St. Lucie	St. Lucie Nuclear Plant
Salem	Salem Nuclear Generating Station
San Onofre	San Onofre Nuclear Generating Station
Seabrook	Seabrook Station
Sequoyah	Sequoyah Nuclear Plant
South Texas	South Texas Project Electric Generating Station
Summer	Virgil C. Summer Nuclear Station
Surry	Surry Power Station
Susquehanna	Susquehanna Steam Electric Station
Three Mile Island	Three Mile Island, Unit 1
Turkey Point	Turkey Point Nuclear Plant
Vermont Yankee	Vermont Yankee Nuclear Power Station
Vogtle	Vogtle Electric Generating Plant
Waterford	Waterford Steam Electric Station
Watts Bar	Watts Bar Nuclear Plant
Wolf Creek	Wolf Creek Generating Station

Units of Measure

ac	acre(s)
bbl	barrel(s)
Btu	British thermal unit(s)
°C	degree(s) Celsius
cm	centimeter(s)
d	day(s)
dB	decibel(s)

°F	degree(s) Fahrenheit
ft	foot (feet)
ft^2	square foot (feet)
ft^3	cubic foot (feet)
gal	gallon(s)
gpd	gallon(s) per day
gpm	gallon(s) per minute
Gy	gray(s)
ha	hectare(s)
hr	hour(s)
Hz	hertz
in.	inch(es)
kg	kilogram(s)
km	kilometer(s)
kV	kilovolt(s)
kW	kilowatt(s)
kWh	kilowatt-hour(s)
L	liter(s)
lb	pound(s)
m	meter(s)
m^2	square meter(s)
m^3	cubic meter(s)
mA	milliampere(s)
mg	milligram(s)
mG	milligauss
mGy	milligray(s)
MHz	megahertz
mi	mile(s)
min	minute(s)
mL	milliliter(s)
MMBtu	million Btu
MPa	megapascal(s)
mph	mile(s) per hour
mrad	milliard(s)
mrem	millirem(s)

Notation

mSv	millisievert(s)
mT	milliTesla(s)
MT	metric tonne(s)
MTHM	metric tonne(s) of heavy metal
MTU	metric tonne(s) of uranium
MW	megawatt(s)
MWe or MW(e)	megawatt(s) electric
MW(t)	megawatt(s) thermal
MWh	megawatt-hour(s)
pCi	picocurie(s)
ppm	part(s) per million
ppmv	parts per million by volume
ppt	part(s) per thousand
psi	pound(s) per square inch
rad	radian
rem	roentgen-equivalent-man
s	second(s)
scf	standard cubic foot (feet)
Sv	sievert(s)
T	tesla(s)
TPY	ton(s) per year
V	volt(s)
yr	year(s)
μCi	microcurie(s)
μGy	microgray(s)
μm	micrometer(s)
μT	microtesla(s)

Conversion Table

Multiply	By	To Obtain
To Convert English to Metric Equivalents		
acres	0.4047	hectares (ha)
cubic feet (ft^3)	0.02832	cubic meters (m^3)
cubic yards (yd^3)	0.7646	cubic meters (m^3)
curies (Ci)	3.7×10^{10}	becquerels (Bq)
degrees Fahrenheit (°F) -32	0.5555	degrees Celsius (°C)
feet (ft)	0.3048	meters (m)
gallons (gal)	3.785	liters (L)
gallons (gal)	0.003785	cubic meters (m^3)
inches (in.)	2.540	centimeters (cm)
miles (mi)	1.609	kilometers (km)
pounds (lb)	0.4536	kilograms (kg)
rads	0.01	grays (Gy)
rems	0.01	sieverts (Sv)
short tons (tons)	907.2	kilograms (kg)
short tons (tons)	0.9072	metric tons (t)
square feet (ft^2)	0.09290	square meters (m^2)
square yards (yd^2)	0.8361	square meters (m^2)
square miles (mi^2)	2.590	square kilometers (km^2)
yards (yd)	0.9144	meters (m)
To Convert Metric to English Equivalents		
becquerels (Bq)	2.7×10^{-11}	curies (Ci)
centimeters (cm)	0.3937	inches (in.)
cubic meters (m^3)	35.31	cubic feet (ft^3)
cubic meters (m^3)	1.308	cubic yards (yd^3)
cubic meters (m^3)	264.2	gallons (gal)
degrees Celsius (°C) +17.78	1.8	degrees Fahrenheit (°F)
grays (Gy)	100	rads
hectares (ha)	2.471	acres
kilograms (kg)	2.205	pounds (lb)
kilograms (kg)	0.001102	short tons (tons)
kilometers (km)	0.6214	miles (mi)
liters (L)	0.2642	gallons (gal)
meters (m)	3.281	feet (ft)
meters (m)	1.094	yards (yd)
metric tons (t)	1.102	short tons (tons)
sieverts (Sv)	100	rems
square kilometers (km^2)	0.3861	square miles (mi^2)
square meters (m^2)	10.76	square feet (ft^2)
square meters (m^2)	1.196	square yards (yd^2)

Appendix B

Comparison of Environmental Issues and Findings in This GEIS Revision to the Issues and Findings in Table B-1 of 10 CFR Part 51 (1996 Version)

Appendix B

Comparison of Environmental Issues and Findings in This GEIS Revision to the Issues and Findings in Table B-1 of 10 CFR Part 51 (1996 Version)

This appendix provides a comparison of the issues and findings presented in this GEIS revision and those issues and findings presented in the 1996 version of Table B-1 of 10 CFR Part 51. For the most part, the 1996 version of Table B-1 of 10 CFR Part 51 reflected the findings of the 1996 GEIS, although a few issues were modified or added after publication of the GEIS (e.g., environmental justice).

Table B-1. Environmental Issues and Findings in This GEIS Revision Compared to the Issues and Findings in Table B-1 of 10 CFR Part 51

New Table B-1		Old Table B-1	
Issue	**Findings**	**Issue**	**Findings**
Land Use			
Onsite land use	**Small (Category 1).** Changes in onsite land use from continued operations and refurbishment associated with license renewal would be a small fraction of the nuclear power plant site and would involve only land that is controlled by the licensee.	Onsite land use	**Small (Category 1).** Projected onsite land use changes required during refurbishment and the renewal period would be a small fraction of any nuclear power plant site and would involve land that is controlled by the applicant.
Offsite land use	**Small (Category 1).** Offsite land use would not be affected by continued operations and refurbishment associated with license renewal.	Offsite land use (refurbishment)	**Small or moderate (Category 2).** Impacts may be of moderate significance at plants in low population areas.
		Offsite land use (license renewal term)	**Small, moderate, or large (Category 2).** Significant changes in land use may be associated with population and tax revenue changes resulting from license renewal.
Offsite land use in transmission line rights-of-way (ROWs)	**Small (Category 1).** Use of transmission line ROWs from continued operations and refurbishment associated with license renewal would continue with no change in land use restrictions.	Power line ROW	**Small (Category 1).** Ongoing use of power line ROWs would continue with no change in restrictions. The effects of these restrictions are of small significance.

Table B-1. (cont.)

New Table B-1		Old Table B-1	
Issue	**Findings**	**Issue**	**Findings**
Visual Resources			
Aesthetic impacts	**Small (Category 1**). No important changes to the visual appearance of plant structures or transmission lines are expected from continued operations and refurbishment associated with license renewal.	Aesthetic impacts (refurbishment)	**Small (Category 1).** No significant impacts are expected during refurbishment.
		Aesthetic impacts (license renewal term)	**Small (Category 1).** No significant impacts are expected during the license renewal term.
		Aesthetic impacts of transmission lines (license renewal term)	**Small (Category 1).** No significant impacts are expected during the license renewal term.
Air Quality			
Air quality impacts (all plants)	**Small (Category 1).** Air quality impacts from continued operations and refurbishment associated with license renewal are expected to be small at all plants. Emissions resulting from refurbishment activities at locations in or near air quality nonattainment or maintenance areas would be short-lived and would cease after these refurbishment activities are completed. Operating experience has shown that the scale of refurbishment activities has not resulted in exceedance of the *de minimis* thresholds for criteria pollutants, and best management practices including fugitive dust controls and the imposition of permit conditions in State and local air emissions permits would ensure	Air quality during refurbishment (nonattainment and maintenance areas)	**Small, moderate, or large (Category 2).** Air quality impacts from plant refurbishment associated with license renewal are expected to be small. However, vehicle exhaust emissions could be cause for concern at locations in or near nonattainment or maintenance areas. The significance of the potential impact cannot be determined without considering the compliance status of each site and the numbers of workers expected to be employed during the outage.

Table B-1. (cont.)

New Table B-1		Old Table B-1	
Issue	**Findings**	**Issue**	**Findings**
Air Quality (cont.)			
Air quality (all plants) (cont.)	conformance with applicable State or Tribal implementation plans. Emissions from emergency diesel generators and fire pumps and routine operations of boilers used for space heating would not be a concern, even for plants located in or adjacent to nonattainment areas. Impacts from cooling tower particulate emissions even under the worst-case situations have been small.		
Air quality effects of transmission lines	**Small (Category 1).** Production of ozone and oxides of nitrogen is insignificant and does not contribute measurably to ambient levels of these gases.	Air quality effects of transmission lines	**Small (Category 1).** Production of ozone and oxides of nitrogen is insignificant and does not contribute measurably to ambient levels of these gases.
Noise			
Noise impacts	**Small (Category 1).** Noise levels would remain below regulatory guidelines for offsite receptors during continued operations and refurbishment associated with license renewal.	Noise	**Small (Category 1).** Noise has not been found to be a problem at operating plants and is not expected to be a problem at any plant during the license renewal term.

Table B-1. (cont.)

New Table B-1		Old Table B-1	
Issue	**Findings**	**Issue**	**Findings**
Geologic Environment			
Geology and soils	**Small (Category 1).** The effect of geologic and soil conditions on plant operations and the impact of continued operations and refurbishment activities on geology and soils would be small for all nuclear power plants and would not change appreciably during the license renewal term.	Not addressed	Not applicable
Surface Water Resources			
Surface water use and quality (non-cooling system impacts)	**Small (Category 1).** Impacts are expected to be small if best management practices are employed to control soil erosion and spills. Surface water use associated with continued operations and refurbishment associated with license renewal would not increase significantly or would be reduced if refurbishment occurs during a plant outage.	Impacts of refurbishment on surface water quality (for all plants) Impacts of refurbishment on surface water use (for all plants)	**Small (Category 1).** Impacts are expected to be negligible during refurbishment because best management practices are expected to be employed to control soil erosion and spills. **Small (Category 1).** Water use during refurbishment will not increase appreciably or will be reduced during plant outage.
Altered current patterns at intake and discharge structures	**Small (Category 1).** Altered current patterns would be limited to the area in the vicinity of the intake and discharge structures. These impacts have been small at operating nuclear power plants.	Altered current patterns at intake and discharge structures (for all plants)	**Small (Category 1).** Altered current patterns have not been found to be a problem at operating nuclear power plants and are not expected to be a problem during the license renewal term.

Table B-1. (cont.)

New Table B-1		Old Table B-1	
Issue	**Findings**	**Issue**	**Findings**
Surface Water Resources (cont.)			
Altered salinity gradients	**Small (Category 1).** Effects on salinity gradients would be limited to the area in the vicinity of the intake and discharge structures. These impacts have been small at operating nuclear power plants.	Altered salinity gradients (for all plants)	**Small (Category 1).** Salinity gradients have not been found to be a problem at operating nuclear power plants and are not expected to be a problem during the license renewal term.
Altered thermal stratification of lakes	**Small (Category 1).** Effects on thermal stratification would be limited to the area in the vicinity of the intake and discharge structures. These impacts have been small at operating nuclear power plants.	Altered thermal stratification of lakes (for all plants)	**Small (Category 1).** Generally, lake stratification has not been found to be a problem at operating nuclear power plants and is not expected to be a problem during the license renewal term.
Scouring caused by discharged cooling water	**Small (Category 1).** Scouring effects would be limited to the area in the vicinity of the intake and discharge structures. These impacts have been small at operating nuclear power plants.	Scouring caused by discharged cooling water (for all plants)	**Small (Category 1).** Scouring has not been found to be a problem at most operating nuclear power plants and has caused only localized effects at a few plants. It is not expected to be a problem during the license renewal term.
Discharge of metals in cooling system effluent	**Small (Category 1).** Discharges of metals have not been found to be a problem at operating nuclear power plants with cooling-tower-based heat dissipation systems and have been satisfactorily mitigated at other plants. Discharges are monitored and controlled as part of the National Pollutant Discharge Elimination System (NPDES) permit process.	Discharge of other metals in waste water (for all plants)	**Small (Category 1).** These discharges have not been found to be a problem at operating nuclear power plants with cooling-tower-based heat dissipation systems and have been satisfactorily mitigated at other plants. They are not expected to be a problem during the license renewal term.

Table B-1. (cont.)

New Table B-1		Old Table B-1	
Issue	**Findings**	**Issue**	**Findings**
Surface Water Resources (cont.)			
Discharge of biocides, sanitary wastes, and minor chemical spills	**Small (Category 1).** The effects of these discharges are regulated by Federal and State environmental agencies. Discharges are monitored and controlled as part of the NPDES permit process. These impacts have been small at operating nuclear power plants.	Discharge of chlorine or other biocides (for all plants)	**Small (Category 1).** Effects are not a concern among regulatory and resource agencies, and are not expected to be a problem during the license renewal term.
		Discharge of sanitary wastes and minor chemical spills (for all plants)	**Small (Category 1).** Effects are readily controlled through NPDES permit rules and periodic modifications, if needed, and are not expected to be a problem during the license renewal term.
Surface water use conflicts (plants with once-through cooling systems)	**Small (Category 1).** These conflicts have not been found to be a problem at operating nuclear power plants with once-through heat dissipation systems.	Water use conflicts (plants with once-through cooling systems)	**Small (Category 1).** These conflicts have not been found to be a problem at operating nuclear power plants with once-through heat dissipation systems.
Surface water use conflicts (plants with cooling ponds or cooling towers using makeup water from a river)	**Small or moderate (Category 2).** Impacts could be of small or moderate significance, depending on makeup water requirements, water availability, and competing water demands.	Water use conflicts (plants with cooling ponds or cooling towers using makeup water from a small river with low flow)	**Small or moderate (Category 2).** The issue has been a concern at nuclear power plants with cooling ponds and at plants with cooling towers. Impacts on instream and riparian communities near these plants could be of moderate significance in some situations.

Table B-1. (cont.)

New Table B-1		Old Table B-1	
Issue	**Findings**	**Issue**	**Findings**
Surface Water Resources (cont.)			
Effects of dredging on surface water quality	**Small (Category 1).** Dredging to remove accumulated sediments in the vicinity of intake and discharge structures and to maintain barge shipping has not been found to be a problem for surface water quality. Dredging is performed under permit from the U.S. Army Corps of Engineers, and possibly, from other State or local agencies.	Not addressed	Not applicable
Temperature effects on sediment transport capacity	**Small (Category 1).** These effects have not been found to be a problem at operating nuclear power plants and are not expected to be a problem.	Temperature effects on sediment transport capacity (for all plants)	**Small (Category 1).** These effects have not been found to be a problem at operating nuclear power plants and are not expected to be a problem during the license renewal term.
Groundwater Resources			
Groundwater contamination and use (non-cooling system impacts)	**Small (Category 1).** Extensive dewatering is not anticipated from continued operations and refurbishment associated with license renewal. Industrial practices involving the use of solvents, hydrocarbons, heavy metals, or other chemicals, and/or the use of wastewater ponds or lagoons have the potential to contaminate site groundwater, soil, and subsoil. Contamination is subject to State or Environmental Protection Agency regulated cleanup and monitoring programs. The application of best	Impacts of refurbishment on groundwater use and quality	**Small (Category 1).** Extensive dewatering during the original construction on some sites will not be repeated during refurbishment on any sites. Any plant wastes produced during refurbishment will be handled in the same manner as in current operating practices and are not expected to be a problem during the license renewal term.

Table B-1. (cont.)

New Table B-1		Old Table B-1	
Issue	**Findings**	**Issue**	**Findings**
Groundwater Resources (cont.)			
Groundwater contamination and use (non-cooling system impacts) (cont.)	management practices for handling any materials produced or used during these activities would reduce impacts.		
Groundwater use conflicts (plants that withdraw less than 100 gallons per minute [gpm])	**Small (Category 1).** Plants that withdraw less than 100 gpm are not expected to cause any groundwater use conflicts.	Groundwater use conflicts (potable and service water; plants that use <100 gpm)	**Small (Category 1).** Plants using less than 100 gpm are not expected to cause any groundwater use conflicts.
Groundwater use conflicts (plants that withdraw more than100 gallons per minute [gpm])	**Small, moderate, or large (Category 2).** Plants that withdraw more than 100 gpm could cause groundwater use conflicts with nearby groundwater users.	Groundwater use conflicts (potable and service water; plants that use >100 gpm)	**Small, moderate, or large (Category 2).** Plants that use more than 100 gpm may cause groundwater use conflicts with nearby groundwater users.
		Groundwater use conflicts (Ranney wells)	**Small, moderate, or large (Category 2).** Ranney wells can result in potential groundwater depression beyond the site boundary. Impacts of large groundwater withdrawal for cooling tower makeup at nuclear power plants using Ranney wells must be evaluated at the time of application for license renewal.

Table B-1. (cont.)

New Table B-1		Old Table B-1	
Issue	**Findings**	**Issue**	**Findings**
Groundwater Resources (cont.)			
Groundwater use conflicts (plants with closed-cycle cooling systems that withdraw makeup water from a river)	**Small, moderate, or large (Category 2).** Water use conflicts could result from water withdrawals from rivers during low-flow conditions, which may affect aquifer recharge. The significance of impacts would depend on makeup water requirements, water availability, and competing water demands.	Groundwater use conflicts (plants using cooling towers withdrawing makeup water from a small river)	**Small, moderate, or large (Category 2).** Water use conflicts may result from surface water withdrawals from small water bodies during low flow conditions which may affect aquifer recharge, especially if other groundwater or upstream surface water users come on line before the time of license renewal.
Groundwater quality degradation resulting from water withdrawals	**Small (Category 1).** Groundwater withdrawals at operating nuclear power plants would not contribute significantly to groundwater quality degradation.	Groundwater quality degradation (Ranney wells)	**Small (Category 1).** Groundwater quality at river sites may be degraded by induced infiltration of poor-quality river water into an aquifer that supplies large quantities of reactor cooling water. However, the lower quality infiltrating water would not preclude the current uses of groundwater and is not expected to be a problem during the license renewal term.
		Groundwater quality degradation (saltwater intrusion)	**Small (Category 1).** Nuclear power plants do not contribute significantly to saltwater intrusion.

Table B-1. (cont.)

New Table B-1		Old Table B-1	
Issue	**Findings**	**Issue**	**Findings**
Groundwater Resources (cont.)			
Groundwater quality degradation (plants with cooling ponds in salt marshes)	**Small (Category 1).** Sites with closed-cycle cooling ponds could degrade groundwater quality. However, groundwater in salt marshes is naturally brackish and thus, not potable. Consequently, the human use of such groundwater is limited to industrial purposes.	Groundwater quality degradation (cooling ponds in salt marshes)	**Small (Category 1).** Sites with closed-cycle cooling ponds may degrade groundwater quality. Because water in salt marshes is brackish, this is not a concern for plants located in salt marshes.
Groundwater quality degradation (plants with cooling ponds at inland sites)	**Small, moderate, or large (Category 2).** Inland sites with closed-cycle cooling ponds could degrade groundwater quality. The significance of the impact would depend on cooling pond water quality, site hydrogeologic conditions (including the interaction of surface water and groundwater), and the location, depth, and pump rate of water wells.	Groundwater quality degradation (cooling ponds at inland sites)	**Small, moderate, or large (Category 2).** Sites with closed-cycle cooling ponds may degrade groundwater quality. For plants located inland, the quality of the groundwater in the vicinity of the ponds must be shown to be adequate to allow continuation of current uses.
Radionuclides released to groundwater	**Small or moderate (Category 2).** Leaks of radioactive liquids from plant components and pipes have occurred at numerous plants. Groundwater protection programs have been established at all operating nuclear power plants to minimize the potential impact from any inadvertent releases. The magnitude of impacts would depend on site-specific characteristics.	Not addressed	Not applicable

Table B-1. (cont.)

New Table B-1		Old Table B-1	
Issue	**Findings**	**Issue**	**Findings**
Terrestrial Resources			
Effects on terrestrial resources (non-cooling system impacts)	**Small, moderate, or large (Category 2).** Impacts resulting from continued operations and refurbishment associated with license renewal may affect terrestrial communities. Application of best management practices would reduce the potential for impacts. The magnitude of impacts would depend on the nature of the activity, the status of the resources that could be affected, and the effectiveness of mitigation.	Refurbishment impacts	**Small, moderate, or large (Category 2).** Refurbishment impacts are insignificant if no loss of important plant and animal habitat occurs. However, it cannot be known whether important plant and animal communities may be affected until the specific proposal is presented with the license renewal application.
Exposure of terrestrial organisms to radionuclides	**Small (Category 1).** Doses to terrestrial organisms from continued operations and refurbishment associated with license renewal are expected to be well below exposure guidelines developed to protect these organisms.	Not addressed	Not applicable

Table B-1. (cont.)

New Table B-1		Old Table B-1	
Issue	**Findings**	**Issue**	**Findings**
Terrestrial Resources (cont.)			
Cooling system impacts on terrestrial resources (plants with once-through cooling systems or cooling ponds)	**Small (Category 1).** No adverse effects to terrestrial plants or animals have been reported as a result of increased water temperatures, fogging, humidity, or reduced habitat quality. Due to the low concentrations of contaminants in cooling system effluents, uptake and accumulation of contaminants in the tissues of wildlife exposed to the contaminated water or aquatic food sources are not expected to be significant issues.	Cooling pond impacts on terrestrial resources	**Small (Category 1).** Impacts of cooling ponds on terrestrial ecological resources are considered to be of small significance at all sites.
Cooling tower impacts on vegetation (plants with cooling towers)	**Small (Category 1).** Impacts from salt drift, icing, fogging, or increased humidity associated with cooling tower operation have the potential to affect adjacent vegetation, but these impacts have been small at operating nuclear power plants and are not expected to change over the license renewal term.	Cooling tower impacts on crops and ornamental vegetation	**Small (Category 1).** Impacts from salt drift, icing, fogging, or increased humidity associated with cooling tower operation have not been found to be a problem at operating nuclear power plants and are not expected to be a problem during the license renewal term.
		Cooling tower impacts on native plants	**Small (Category 1).** Impacts from salt drift, icing, fogging, or increased humidity associated with cooling tower operation have not been found to be a problem at operating nuclear power plants and are not expected to be a problem during the license renewal term.

Table B-1. (cont.)

New Table B-1		Old Table B-1	
Issue	**Findings**	**Issue**	**Findings**
Terrestrial Resources (cont.)			
Bird collisions with plant structures and transmission lines	**Small (Category 1).** Bird collisions with cooling towers and other plant structures and transmission lines occur at rates that are unlikely to affect local or migratory populations and the rates are not expected to change.	Bird collisions with cooling towers	**Small (Category 1).** These collisions have not been found to be a problem at operating nuclear power plants and are not expected to be a problem during the license renewal term.
		Bird collisions with power lines	**Small (Category 1).** Impacts are expected to be of small significance at all sites.
Water use conflicts with terrestrial resources (plants with cooling ponds or cooling towers using makeup water from a river)	**Small or moderate (Category 2).** Impacts on terrestrial resources in riparian communities affected by water use conflicts could be of moderate significance.	Water use conflicts (plants with cooling ponds or cooling towers using makeup water from a small river with low flow)	**Small or moderate (Category 2).** The issue has been a concern at nuclear power plants with cooling ponds and at plants with cooling towers. Impacts on instream and riparian communities near these plants could be of moderate significance in some situations.
Transmission line ROW management impacts on terrestrial resources	**Small (Category 1).** Continued ROW management during the license renewal term is expected to keep terrestrial communities in their current condition. Application of best management practices would reduce the potential for impacts.	Power line ROW management (cutting and herbicide application)	**Small (Category 1).** The impacts of ROW maintenance on wildlife are expected to be of small significance at all sites.
		Floodplains and wetland on power line ROW	**Small (Category 1).** Periodic vegetation control is necessary in forested wetlands underneath power lines and can be achieved with minimal damage to the wetland. No significant impact is expected at any nuclear power plant during the license renewal term.

Table B-1. (cont.)

New Table B-1		Old Table B-1	
Issue	**Findings**	**Issue**	**Findings**
Terrestrial Resources (cont.)			
Electromagnetic fields on flora and fauna (plants, agricultural crops, honeybees, wildlife, livestock)	**Small (Category 1).** No significant impacts of electromagnetic fields on terrestrial flora and fauna have been identified. Such effects are not expected to be a problem during the license renewal term.	Impacts of electromagnetic fields on flora and fauna (plants, agricultural crops, honeybees, wildlife, livestock)	**Small (Category 1).** No significant impacts of electromagnetic fields on terrestrial flora and fauna have been identified. Such effects are not expected to be a problem during the license renewal term.
Aquatic Resources			
Impingement and entrainment of aquatic organisms (plants with once-through cooling systems or cooling ponds)	**Small, moderate, or large (Category 2).** The impacts of impingement and entrainment are small at many plants but may be moderate or even large at a few plants with once-through and cooling-pond cooling systems, depending on cooling system withdrawal rates and volumes and the aquatic resources at the site.	Impingement of fish and shellfish (for plants with once-through and cooling-pond heat dissipation systems)	**Small, moderate, or large (Category 2).** The impacts of impingement are small at many plants but may be moderate or even large at a few plants with once-through and cooling-pond cooling systems.
		Entrainment of fish and shellfish in early life stages (for plants with once-through and cooling-pond heat dissipation systems)	**Small, moderate, or large (Category 2).** The impacts of entrainment are small at many plants but may be moderate or even large at a few plants with once-through and cooling-pond cooling systems. Further, ongoing efforts in the vicinity of these plants to restore fish populations may increase the numbers of fish susceptible to intake effects during the license renewal period, such that entrainment studies conducted in support of the original license may no longer be valid.

Table B-1. (cont.)

New Table B-1		Old Table B-1	
Issue	**Findings**	**Issue**	**Findings**
Aquatic Resources (cont.)			
Impingement and entrainment of aquatic organisms (plants with cooling towers)	**Small (Category 1).** Impingement and entrainment rates are lower at plants that use closed-cycle cooling with cooling towers because the rates and volumes of water withdrawal needed for makeup are minimized.	Impingement of fish and shellfish (for plants with cooling-tower-based heat dissipation systems)	**Small (Category 1).** The impingement has not been found to be a problem at operating nuclear power plants with this type of cooling system and is not expected to be a problem during the license renewal term.
		Entrainment of fish and shellfish in early life stages (for plants with cooling-tower-based heat dissipation systems)	**Small (Category 1).** Entrainment of fish has not been found to be a problem at operating nuclear power plants with this type of cooling system and is not expected to be a problem during the license renewal term.
Entrainment of phytoplankton and zooplankton (all plants)	**Small (Category 1).** Entrainment of phytoplankton and zooplankton has not been found to be a problem at operating nuclear power plants and is not expected to be a problem during the license renewal term.	Entrainment of phytoplankton and zooplankton (for all plants)	**Small (Category 1).** Entrainment of phytoplankton and zooplankton has not been found to be a problem at operating nuclear power plants and is not expected to be a problem during the license renewal term.
Thermal impacts on aquatic organisms (plants with once-through cooling systems or cooling ponds)	**Small, moderate, or large (Category 2).** Most of the effects associated with thermal discharges are localized and are not expected to affect overall stability of populations or resources. The magnitude of impacts, however, would depend on site-specific thermal plume characteristics and the nature of aquatic resources in the area.	Heat shock (for plants with once-through and cooling-pond heat dissipation systems)	**Small, moderate, or large (Category 2).** Because of continuing concerns about heat shock and the possible need to modify thermal discharges in response to changing environmental conditions, the impacts may be of moderate or large significance at some plants.

Table B-1. (cont.)

New Table B-1		Old Table B-1	
Issue	**Findings**	**Issue**	**Findings**
Aquatic Resources (cont.)			
Thermal impacts on aquatic organisms (plants with cooling towers)	**Small (Category 1).** Thermal effects associated with plants that use cooling towers are expected to be small because of the reduced amount of heated discharge.	Heat shock (for plants with cooling-tower-based heat dissipation systems)	**Small (Category 1).** Heat shock has not been found to be a problem at operating nuclear power plants with this type of cooling system and is not expected to be a problem during the license renewal term.
Infrequently reported thermal impacts (all plants)	**Small (Category 1).** Continued operations during the license renewal term are expected to have small thermal impacts with respect to the following: Cold shock has been satisfactorily mitigated at operating nuclear plants with once-through cooling systems, has not endangered fish populations or been found to be a problem at operating nuclear power plants with cooling towers or cooling ponds, and is not expected to be a problem.	Cold shock (for all plants)	**Small (Category 1).** Cold shock has been satisfactorily mitigated at operating nuclear plants with once-through cooling systems, has not endangered fish populations or been found to be a problem at operating nuclear power plants with cooling towers or cooling ponds, and is not expected to be a problem during the license renewal term.
	Thermal plumes have not been found to be a problem at operating nuclear power plants and are not expected to be a problem.	Thermal plume barrier to migrating fish (for all plants)	**Small (Category 1).** Thermal plumes have not been found to be a problem at operating nuclear power plants and are not expected to be a problem during the license renewal term.
	Thermal discharge may have localized effects but is not expected to affect the larger geographical distribution of aquatic organisms.	Distribution of aquatic organisms (for all plants)	**Small (Category 1).** Thermal discharge may have localized effects but is not expected to affect the larger geographical distribution of aquatic organisms.

Table B-1. (cont.)

New Table B-1		Old Table B-1	
Issue	**Findings**	**Issue**	**Findings**
Aquatic Resources (cont.)			
Infrequently reported thermal impacts (all plants) (cont.)	Premature emergence has been found to be a localized effect at some operating nuclear power plants but has not been a problem and is not expected to be a problem.	Premature emergence of aquatic insects (for all plants)	**Small (Category 1).** Premature emergence has been found to be a localized effect at some operating nuclear power plants but has not been a problem and is not expected to be a problem during the license renewal term.
	Stimulation of nuisance organisms has been satisfactorily mitigated at the single nuclear power plant with a once-through cooling system where previously it was a problem. It has not been found to be a problem at operating nuclear power plants with cooling towers or cooling ponds and is not expected to be a problem.	Stimulation of nuisance organisms (e.g., shipworms)	**Small (Category 1).** Stimulation of nuisance organisms has been satisfactorily mitigated at the single nuclear power plant with a once-through cooling system where previously it was a problem. It has not been found to be a problem at operating nuclear power plants with cooling towers or cooling ponds and is not expected to be a problem during the license renewal term.

Table B-1. (cont.)

New Table B-1		Old Table B-1	
Issue	**Findings**	**Issue**	**Findings**
Aquatic Resources (cont.)			
Effects of cooling water discharge on dissolved oxygen, gas supersaturation, and eutrophication	**Small (Category 1).** Gas supersaturation was a concern at a small number of operating nuclear power plants with once-through cooling systems but has been mitigated. Low dissolved oxygen was a concern at one nuclear power plant with a once-through cooling system but has been mitigated. Eutrophication (nutrient loading) and resulting effects on chemical and biological oxygen demands have not been found to be a problem at operating nuclear power plants.	Gas supersaturation (gas bubble disease) (for all plants)	**Small (Category 1).** Gas supersaturation was a concern at a small number of operating nuclear power plants with once-through cooling systems but has been satisfactorily mitigated. It has not been found to be a problem at operating nuclear power plants with cooling towers or cooling ponds and is not expected to be a problem during the license renewal term.
		Low dissolved oxygen in the discharge (for all plants)	**Small (Category 1).** Low dissolved oxygen has been a concern at one nuclear power plant with a once-through cooling system but has been effectively mitigated. It has not been found to be a problem at operating nuclear power plants with cooling towers or cooling ponds and is not expected to be a problem during the license renewal term.
		Eutrophication (for all plants)	**Small (Category 1).** Eutrophication has not been found to be a problem at operating nuclear power plants and is not expected to be a problem during the license renewal term.

Table B-1. (cont.)

New Table B-1		Old Table B-1	
Issue	**Findings**	**Issue**	**Findings**
Aquatic Resources (cont.)			
Effects of nonradiological contaminants on aquatic organisms	**Small (Category 1).** Best management practices and discharge limitations of NPDES permits are expected to minimize the potential for impacts to aquatic resources during continued operations and refurbishment associated with license renewal. Accumulation of metal contaminants has been a concern at a few nuclear power plants but has been satisfactorily mitigated by replacing copper alloy condenser tubes with those of another metal.	Accumulation of contaminants in sediments or biota (for all plants)	**Small (Category 1).** Accumulation of contaminants has been a concern at a few nuclear power plants but has been satisfactorily mitigated by replacing copper alloy condenser tubes with those of another metal. It is not expected to be a problem during the license renewal term.
Exposure of aquatic organisms to radionuclides	**Small (Category 1).** Doses to aquatic organisms are expected to be well below exposure guidelines developed to protect these aquatic organisms.	Not addressed	Not applicable
Effects of dredging on aquatic resources	**Small (Category 1).** Dredging at nuclear power plants is expected to occur infrequently, would be of relatively short duration, and would affect relatively small areas. Dredging is performed under permit from the U.S. Army Corps of Engineers, and possibly from other State or local agencies.	Not addressed	Not applicable

Table B-1. (cont.)

New Table B-1		Old Table B-1	
Issue	**Findings**	**Issue**	**Findings**
Aquatic Resources (cont.)			
Water use conflicts with aquatic resources (plants with cooling ponds or cooling towers using makeup water from a river)	**Small or moderate (Category 2).** Impacts on aquatic resources in stream communities affected by water use conflicts could be of moderate significance in some situations.	Water use conflicts (plants with cooling ponds or cooling towers using makeup water from a small river with low flow)	**Small or moderate (Category 2).** The issue has been a concern at nuclear power plants with cooling ponds and at plants with cooling towers. Impacts on instream and riparian communities near these plants could be of moderate significance in some situations.
Effects on aquatic resources (non-cooling system impacts)	**Small (Category 1).** Licensee application of appropriate mitigation measures is expected to result in no more than small changes to aquatic communities from their current condition.	Refurbishment (for all plants)	**Small (Category 1).** During plant shutdown and refurbishment there will be negligible effects on aquatic biota because of a reduction of entrainment and impingement of organisms or a reduced release of chemicals.
Impacts of transmission line ROW management on aquatic resources	**Small (Category 1).** Licensee application of best management practices to ROW maintenance is expected to result in no more than small impacts to aquatic resources.	Not addressed	Not applicable
Losses from predation, parasitism, and disease among organisms exposed to sublethal stresses	**Small (Category 1).** These types of losses have not been found to be a problem at operating nuclear power plants and are not expected to be a problem during the license renewal term.	Losses from predation, parasitism, and disease among organisms exposed to sublethal stresses (for all plants)	**Small (Category 1).** These types of losses have not been found to be a problem at operating nuclear power plants and are not expected to be a problem during the license renewal term.

Table B-1. (cont.)

New Table B-1		Old Table B-1	
Issue	**Findings**	**Issue**	**Findings**
Special Status Species and Habitats			
Threatened, endangered, and protected species, critical habitat and essential fish habitat	**(Category 2).** The magnitude of impacts on threatened, endangered, and protected species, critical habitat, and essential fish habitat would depend on the occurrence of listed species and habitats and the effects of power plant systems on them. Consultation with appropriate agencies would be needed to determine whether special status species or habitats are present and whether they would be adversely affected by continued operations and refurbishment associated with license renewal.	Threatened or endangered species (for all plants)	**Small, moderate, or large (Category 2).** Generally, plant refurbishment and continued operation are not expected to adversely affect threatened or endangered species. However, consultation with appropriate agencies would be needed at the time of license renewal to determine whether threatened or endangered species are present and whether they would be adversely affected.

Table B-1. (cont.)

New Table B-1		Old Table B-1	
Issue	**Findings**	**Issue**	**Findings**
Historic and Cultural Resources			
Historic and Cultural resources	**(Category 2).** Continued operations and refurbishment associated with license renewal are expected to have no more than small impacts on historic and cultural resources located onsite and in the transmission line ROW because most impacts could be mitigated by avoiding those resources. The National Historic Preservation Act (NHPA) requires the Federal agency to consult with the State Historic Preservation Officer (SHPO) and appropriate Native American Tribes to determine the potential effects on historic properties and mitigation, if necessary.	Historic and archaeological resources	**Small, moderate, or large (Category 2).** Generally, plant refurbishment and continued operation are expected to have no more than small adverse impacts on historic and archaeological resources. However, the National Historic Preservation Act requires the Federal agency to consult with the State Historic Preservation Officer to determine whether there are properties present that require protection.
Socioeconomics			
Employment and income, recreation and tourism	**Small (Category 1).** Although most nuclear plants have large numbers of employees with higher than average wages and salaries, employment, income, recreation, and tourism, impacts from continued operations and refurbishment associated with license renewal are expected to be small.	Public services: public safety, social services, and tourism and recreation	**Small (Category 1).** Impacts to public safety, social services, and tourism and recreation are expected to be of small significance at all sites.

Table B-1. (cont.)

New Table B-1		Old Table B-1	
Issue	**Findings**	**Issue**	**Findings**
Socioeconomics (cont.)			
Tax revenues	**Small (Category 1).** Nuclear plants provide tax revenue to local jurisdictions in the form of property tax payments, payments in lieu of tax (PILOT), or tax payments on energy production. The amount of tax revenue paid during the license renewal term as a result of continued operations and refurbishment associated with license renewal is not expected to change.	Considered in the 1996 GEIS, but not identified as an issue	Not applicable
Community services and education	**Small (Category 1).** Changes resulting from continued operations and refurbishment associated with license renewal to local community and educational services would be small. With little or no change in employment at the licensee's plant, value of the power plant, payments on energy production, and PILOT payments expected during the license renewal term, community and educational services would not be affected by continued power plant operations.	Public services: public safety, social services, and tourism and recreation	**Small (Category 1).** Impacts to public safety, social services, and tourism and recreation are expected to be of small significance at all sites.
		Public services: public utilities	**Small or moderate (Category 2).** An increased problem with water shortages at some sites may lead to impacts of moderate significance on public water supply availability.
		Public services, education (license renewal term)	**Small (Category 1).** Only impacts of small significance are expected.
		Public services, education (refurbishment)	**Small, moderate, or large (Category 2).** Most sites would experience impacts of small significance but larger impacts are possible depending on site- and project-specific factors.

Table B-1. (cont.)

New Table B-1		Old Table B-1	
Issue	**Findings**	**Issue**	**Findings**
Socioeconomics (cont.)			
Population and housing	**Small (Category 1).** Changes resulting from continued operations and refurbishment associated with license renewal to regional population and housing availability and value would be small. With little or no change in employment at the licensee's plant expected during the license renewal term, population and housing availability and values would not be affected by continued power plant operations.	Housing impacts	**Small, moderate, or large (Category 2).** Housing impacts are expected to be of small significance at plants located in a medium or high population area and not in an area where growth control measures that limit housing development are in effect. Moderate or large housing impacts of the workforce associated with refurbishment may be associated with plants located in sparsely populated areas or in areas with growth control measures that limit housing development.
Transportation	**Small (Category 1).** Changes resulting from continued operations and refurbishment associated with license renewal to traffic volumes would be small.	Public services, transportation	**Small, moderate, or large (Category 2).** Transportation impacts (level of service) of highway traffic generated during plant refurbishment and during the term of the renewed license are generally expected to be of small significance. However, the increase in traffic associated with the additional workers and the local road and traffic control conditions may lead to impacts of moderate or large significance at some sites.

Table B-1. (cont.)

New Table B-1		Old Table B-1	
Issue	**Findings**	**Issue**	**Findings**
Human Health			
Radiation exposures to the public	**Small (Category 1).** Radiation doses to the public from continued operations and refurbishment associated with license renewal are expected to continue at current levels, and would be well below regulatory limits.	Radiation exposures to the public during refurbishment	**Small (Category 1).** During refurbishment, the gaseous effluents would result in doses that are similar to those from current operation. Applicable regulatory dose limits to the public are not expected to be exceeded.
		Radiation exposures to the public (license renewal term)	**Small (Category 1).** Radiation doses to the public will continue at current levels associated with normal operations.
Radiation exposures to plant workers	**Small (Category 1).** Occupational doses from continued operations and refurbishment associated with license renewal are expected to be within the range of doses experienced during the current license term, and would continue to be well below regulatory limits.	Occupational radiation exposures during refurbishment	**Small (Category 1).** Occupational doses from refurbishment are expected to be within the range of annual average collective doses experienced for pressurized-water reactors and boiling-water reactors. Occupational mortality risk from all causes, including radiation, is in the mid-range for industrial settings.
		Occupational radiation exposures (license renewal term)	**Small (Category 1).** Projected maximum occupational doses during the license renewal term are within the range of doses experienced during normal operations and normal maintenance outages, and would be well below regulatory limits.

Table B-1. (cont.)

New Table B-1		Old Table B-1	
Issue	**Findings**	**Issue**	**Findings**
Human Health (cont.)			
Human health impact from chemicals	**Small (Category 1).** Chemical hazards to plant workers resulting from continued operations and refurbishment associated with license renewal are expected to be minimized by the licensee implementing good industrial hygiene practices as required by permits and Federal and State regulations. Chemical releases to the environment and the potential for impacts to the public are expected to be minimized by adherence to discharge limitations of NPDES and other permits.	Not addressed	Not applicable
Microbiological hazards to the public (plants with cooling ponds or canals or cooling towers that discharge to a river)	**Small, moderate, or large (Category 2).** These organisms are not expected to be a problem at most operating plants except possibly at plants using cooling ponds, lakes, or canals that discharge into rivers. Impacts would depend on site-specific characteristics.	Microbiological organisms (public health) (plants using lakes or canals, or cooling towers or cooling ponds that discharge to a small river)	**Small, moderate, or large (Category 2).** These organisms are not expected to be a problem at most operating plants except possibly at plants using cooling ponds, lakes, or canals that discharge to small rivers. Without site-specific data, it is not possible to predict the effects generically.
Microbiological hazards to plant workers	**Small (Category 1).** Occupational health impacts are expected to be controlled by continued application of accepted industrial hygiene practices to minimize worker exposures as required by permits and Federal and State regulations.	Microbiological organisms (occupational health)	**Small (Category 1).** Occupational health impacts are expected to be controlled by continued application of accepted industrial hygiene practices to minimize worker exposures.

Table B-1. (cont.)

New Table B-1		Old Table B-1	
Issue	**Findings**	**Issue**	**Findings**
Human Health (cont.)			
Chronic effects of electromagnetic fields (EMFs)	**Uncertain.** Studies of 60-Hz EMFs have not uncovered consistent evidence linking harmful effects with field exposures. EMFs are unlike other agents that have a toxic effect (e.g., toxic chemicals and ionizing radiation) in that dramatic acute effects cannot be forced and longer-term effects, if real, are subtle. Because the state of the science is currently inadequate, no generic conclusion on human health impacts is possible.	Electromagnetic fields, chronic effects	**Uncertain.** Biological and physical studies of 60-Hz electromagnetic fields have not found consistent evidence linking harmful effects with field exposures. However, research is continuing in this area and a consensus scientific view has not been reached.
Physical occupational hazards	**Small (Category 1).** Occupational safety and health hazards are generic to all types of electrical generating stations, including nuclear power plants, and are of small significance if the workers adhere to safety standards and use protective equipment as required by Federal and State regulations.	Not addressed	Not applicable

Table B-1. (cont.)

New Table B-1		Old Table B-1	
Issue	**Findings**	**Issue**	**Findings**
Human Health (cont.)			
Electric shock hazards	**Small, moderate, or large (Category 2).** Electrical shock potential is of small significance for transmission lines that are operated in adherence with the National Electrical Safety Code (NESC). Without a review of conformance with NESC criteria of each nuclear plant's in-scope transmission lines, it is not possible to determine the significance of the electrical shock potential.	Electromagnetic fields, acute effects (electric shock)	**Small, moderate, or large (Category 2).** Electrical shock resulting from direct access to energized conductors or from induced charges in metallic structures have not been found to be a problem at most operating plants and generally are not expected to be a problem during the license renewal term. However, site-specific review is required to determine the significance of the electric shock potential at the site.
Postulated Accidents			
Design-basis accidents	**Small (Category 1).** The NRC staff has concluded that the environmental impacts of design-basis accidents are of small significance for all plants.	Design basis accidents	**Small (Category 1).** The NRC staff has concluded that the environmental impacts of design basis accidents are of small significance for all plants.

Table B-1. (cont.)

New Table B-1		Old Table B-1	
Issue	**Findings**	**Issue**	**Findings**
Postulated Accidents (cont.)			
Severe accidents	**Small (Category 2).** The probability-weighted consequences of atmospheric releases, fallout onto open bodies of water, releases to groundwater, and societal and economic impacts from severe accidents are small for all plants. However, alternatives to mitigate severe accidents must be considered for all plants that have not considered such alternatives.	Severe accidents	**Small (Category 2).** The probability-weighted consequences of atmospheric releases, fallout onto open bodies of water, releases to ground water, and societal and economic impacts from severe accidents are small for all plants. However, alternatives to mitigate severe accidents must be considered for all plants that have not considered such alternatives.
Environmental Justice			
Minority and low-income populations	**(Category 2).** Impacts to minority and low-income populations and subsistence consumption resulting from continued operations and refurbishment associated with license renewal will be addressed in plant-specific reviews. See NRC Policy Statement on the Treatment of Environmental Justice Matters in NRC Regulatory and Licensing Actions (69 FR 52040, August 24, 2004).	Environmental justice	**None.** The need for and the content of an analysis of environmental justice will be addressed in plant-specific reviews.

Table B-1. (cont.)

New Table B-1		Old Table B-1	
Issue	**Findings**	**Issue**	**Findings**
Solid Waste Management			
Low-level waste storage and disposal	**Small (Category 1).** The comprehensive regulatory controls that are in place and the low public doses being achieved at reactors ensure that the radiological impacts to the environment would remain small during the license renewal term.	Low-level waste storage and disposal	**Small (Category 1).** The comprehensive regulatory controls that are in place and the low public doses being achieved at reactors ensure that the radiological impacts to the environment will remain small during the term of a renewed license. The maximum additional onsite land that may be required for low-level waste storage during the term of a renewed license and associated impacts will be small. Nonradiological impacts on air and water will be negligible. The radiological and nonradiological environmental impacts of long-term disposal of low-level waste from any individual plant at licensed sites are small. In addition, the Commission concludes that there is reasonable assurance that sufficient low-level waste disposal capacity will be made available when needed for facilities to be decommissioned consistent with NRC decommissioning requirements.

Table B-1. (cont.)

New Table B-1		Old Table B-1	
Issue	**Findings**	**Issue**	**Findings**
Solid Waste Management (cont.)			
Onsite storage of spent nuclear fuel	**Small (Category 1).** The expected increase in the volume of spent fuel from an additional 20 years of operation can be safely accommodated onsite during the license renewal term with small environmental effects through dry or pool storage at all plants.	Onsite spent fuel	**Small (Category 1).** The expected increase in the volume of spent fuel from an additional 20 years of operation can be safely accommodated onsite with small environmental effects through dry or pool storage at all plants if a permanent repository or monitored retrievable storage is not available.
Offsite radiological impacts of spent nuclear fuel and high-level waste disposal	**Uncertain impact.** The generic conclusion on offsite radiological impacts of spent nuclear fuel and high-level waste is not being finalized pending the completion of a generic environmental impact statement on waste confidence.[a]	Offsite radiological impacts (spent fuel and high-level waste disposal)	The NRC did not assign a single level of significance for the impacts of spent fuel and high-level waste disposal, but considered the issue Category 1.[b]

Table B-1. (cont.)

New Table B-1		Old Table B-1	
Issue	**Findings**	**Issue**	**Findings**
Solid Waste Management (cont.)			
Mixed waste storage and disposal	**Small (Category 1).** The comprehensive regulatory controls and the facilities and procedures that are in place ensure proper handling and storage, as well as negligible doses and exposure to toxic materials for the public and the environment at all plants. License renewal would not increase the small, continuing risk to human health and the environment posed by mixed waste at all plants. The radiological and nonradiological environmental impacts of long-term disposal of mixed waste from any individual plant at licensed sites are small.	Mixed waste storage and disposal	**Small (Category 1).** The comprehensive regulatory controls and the facilities and procedures that are in place ensure proper handling and storage, as well as negligible doses and exposure to toxic materials for the public and the environment at all plants. License renewal will not increase the small, continuing risk to human health and the environment posed by mixed waste at all plants. The radiological and nonradiological environmental impacts of long-term disposal of mixed waste from any individual plant at licensed sites are small. In addition, the Commission concludes that there is reasonable assurance that sufficient mixed waste disposal capacity will be made available when needed for facilities to be decommissioned consistent with NRC decommissioning requirements.

Table B-1. (cont.)

New Table B-1		Old Table B-1	
Issue	**Findings**	**Issue**	**Findings**
Solid Waste Management (cont.)			
Nonradioactive waste storage and disposal	**Small (Category 1).** No changes to systems that generate nonradioactive waste are anticipated during the license renewal term. Facilities and procedures are in place to ensure continued proper handling, storage, and disposal, as well as negligible exposure to toxic materials for the public and the environment at all plants.	Nonradiological waste	**Small (Category 1).** No changes to generating systems are anticipated for license renewal. Facilities and procedures are in place to ensure continued proper handling and disposal at all plants.
Cumulative Impacts			
Cumulative impacts	**(Category 2).** Cumulative impacts of continued operations and refurbishment associated with license renewal must be considered on a plant-specific basis. Impacts would depend on regional resource characteristics, the resource-specific impacts of license renewal, and the cumulative significance of other factors affecting the resource.	Not addressed	Not applicable
Uranium Fuel Cycle			
Offsite radiological impacts—individual impacts from other than the disposal of spent fuel and high-level waste	**Small (Category 1).** The impacts to the public from radiological exposures have been considered by the Commission in Table S-3 of this part. Based on information in the GEIS, impacts to individuals from radioactive gaseous and liquid releases, including radon-222 and technetium-99, would remain at or below the NRC's regulatory limits.	Offsite radiological impacts (individual effects from other than the disposal of spent fuel and high-level waste	**Small (Category 1).** Offsite impacts of the uranium fuel cycle have been considered by the Commission in Table S-3 of this part. Based on information in the GEIS, impacts on individuals from radioactive gaseous and liquid releases including radon-222 and technetium-99 are small.

Table B-1. (cont.)

New Table B-1		Old Table B-1	
Issue	**Findings**	**Issue**	**Findings**
Uranium Fuel Cycle (cont.)			
Offsite radiological impacts—collective impacts from other than the disposal of spent fuel and high-level waste	**(Category 1).** There are no regulatory limits applicable to collective doses to the general public from fuel-cycle facilities. The practice of estimating health effects on the basis of collective doses may not be meaningful. All fuel-cycle facilities are designed and operated to meet the applicable regulatory limits and standards. The Commission concludes that the collective impacts are acceptable. The Commission concludes that the impacts would not be sufficiently large to require the NEPA conclusion, for any plant, that the option of extended operation under 10 CFR Part 54 should be eliminated. Accordingly, while the Commission has not assigned a single level of significance for the collective impacts of the uranium fuel cycle, this issue is considered Category 1.	Offsite radiological impacts (collective effects)	The NRC did not assign a single level of significance for the collective effects of the fuel cycle, but considered the issue Category 1.[(c)]
Nonradiological impacts of the uranium fuel cycle	**Small (Category 1).** The nonradiological impacts of the uranium fuel cycle resulting from the renewal of an operating license for any plant would be small.	Nonradiological impacts of the uranium fuel cycle	**Small (Category 1).** The nonradiological impacts of the uranium fuel cycle resulting from the renewal of an operating license for any plant are found to be small.

Table B-1. (cont.)

New Table B-1		Old Table B-1	
Issue	**Findings**	**Issue**	**Findings**
Uranium Fuel Cycle (cont.)			
Transportation	**Small (Category 1).** The impacts of transporting materials to and from uranium-fuel-cycle facilities on workers, the public, and the environment are expected to be small.	Transportation	**Small (Category 1).** The impacts of transporting spent fuel enriched up to 5 percent uranium-235, with average burnup for the peak rod, to current levels approved by NRC up to 62,000 MWd/MTU and the cumulative impacts of transporting high-level waste to a single repository, such as Yucca Mountain, Nevada, are found to be consistent with the impact values contained in 10 CFR 51.52(c), Summary Table S–4—Environmental Impact of Transportation of Fuel and Waste to and from One Light-Water-Cooled Nuclear Power Reactor. If fuel enrichment or burnup conditions are not met, the applicant must submit an assessment of the implications for the environmental impact values reported in § 51.52.

Table B-1. (cont.)

New Table B-1		Old Table B-1	
Issue	**Findings**	**Issue**	**Findings**
Termination of Nuclear Power Plant Operations and Decommissioning			
Termination of plant operations and decommissioning	**Small (Category 1).** License renewal is expected to have a negligible effect on the impacts of terminating operations and decommissioning on all resources.	Air quality	**Small (Category 1).** Air quality impacts of decommissioning are expected to be negligible either at the end of the current operating term or at the end of the license renewal term.
		Water quality	**Small (Category 1).** The potential for significant water quality impacts from erosion or spills is no greater whether decommissioning occurs after a 20-year license renewal period or after the original 40-year operation period, and measures are readily available to avoid such impacts.
		Ecological resources	**Small (Category 1).** Decommissioning after either the initial operating period or after a 20-year license renewal period is not expected to have any direct ecological impacts.

Table B-1. (cont.)

New Table B-1		Old Table B-1	
Issue	**Findings**	**Issue**	**Findings**
Termination of Nuclear Power Plant Operations and Decommissioning (cont.)			
Termination of plant operations and decommissioning (cont.)		Socioeconomic impacts	**Small (Category 1).** Decommissioning would have some short-term socioeconomic impacts. The impacts would not be increased by delaying decommissioning until the end of a 20-year relicense period, but they might be decreased by population and economic growth.
		Radiation doses	**Small (Category 1).** Doses to the public will be well below applicable regulatory standards regardless of which decommissioning method is used. Occupational doses would increase no more than 1 man-rem caused by buildup of long-lived radionuclides during the license renewal term.
		Waste management	**Small (Category 1).** Decommissioning at the end of a 20-year license renewal period would generate no more solid wastes than at the end of the current license term. No increase in the quantities of Class C or greater than Class C wastes would be expected.

Table B-1. (cont.)

(a) As a result of the decision of United States Court of Appeals in *New York v. NRC*, 681 F.3d 471 (D.C. Cir. 2012), the NRC cannot rely upon its waste confidence decision and rule until it has taken those actions that will address the deficiencies identified by the D.C. Circuit. Although the waste confidence decision and rule did not assess the impacts associated with disposal of spent nuclear fuel and high-level waste in a repository, it did reflect the Commission's confidence, at the time, in the technical feasibility of a repository and when that repository could have been expected to become available. Without the analysis in the waste confidence decision and rule regarding the technical feasibility and availability of a repository, the NRC cannot assess how long the spent fuel will need to be stored onsite.

(b) For the high-level waste and spent fuel disposal component of the fuel cycle, there are no current regulatory limits for offsite releases of radionuclides for the current candidate repository site. However, if we assume that limits are developed along the lines of the 1995 National Academy of Sciences (NAS) report, *Technical Bases for Yucca Mountain Standards*, and that in accordance with the Commission's Waste Confidence Decision, 10 CFR 51.23, a repository can and likely will be developed at some site that will comply with such limits, peak doses to virtually all individuals will be 100 millirem per year or less. However, while the Commission has reasonable confidence that these assumptions will prove correct, there is considerable uncertainty since the limits are yet to be developed, no repository application has been completed or reviewed, and uncertainty is inherent in the models used to evaluate possible pathways to the human environment. The NAS report indicated that 100 millirem per year should be considered as a starting point for limits for individual doses, but notes that some measure of consensus exists among national and international bodies that the limits should be a fraction of the 100 millirem per year. The lifetime individual risk from 100 millirem annual dose limit is about 3×10^{-3}.

Estimating cumulative doses to populations over thousands of years is more problematic. The likelihood and consequences of events that could seriously compromise the integrity of a deep geologic repository were evaluated by the Department of Energy in the *Final Environmental Impact Statement: Management of Commercially Generated Radioactive Waste*, October 1980. The evaluation estimated the 70-year whole-body dose commitment to the maximally exposed individual (MEI) and to the regional population resulting from several modes of breaching a reference repository in the year of closure, after 1,000 years, after 100,000 years, and after 100,000,000 years. Subsequently, the NRC and other Federal agencies have expended considerable effort to develop models for the design and for the licensing of a high-level waste repository, especially for the candidate repository at Yucca Mountain. More meaningful estimates of doses to the population may be poss ble in the future as more is understood about the performance of the proposed Yucca Mountain repository. Such estimates would involve very great uncertainty, especially with respect to cumulative population doses over thousands of years. The standard proposed by the NAS is a limit on maximum individual dose. The relationship of potential new regulatory requirements, based on the NAS report, and cumulative population impacts have not been determined, although the report articulates the view that protection of individuals will adequately protect the population for a repository at Yucca Mountain. However, the EPA's generic repository standards in 40 CFR Part 191 generally provide an indication of the order of magnitude of cumulative risk to population that could result from the licensing of a Yucca Mountain repository, assuming the ultimate standards will be within the range of standards now under consideration. The standards in 40 CFR Part 191 protect the population by imposing limitations on the amount of radioactive material released over 10,000 years. The cumulative release limits are based on EPA's population impact goal of 1,000 premature cancer deaths worldwide for a 100,000 metric tonne (MTHM) repository.

Nevertheless, despite all the uncertainty, some judgment as to the regulatory NEPA implications of these matters should be made, and it makes no sense to repeat the same judgment in every case. Even taking the uncertainties into account, the Commission concludes that these impacts are acceptable in that these impacts would not be sufficiently large to require the NEPA conclusion, for any plant, that the option of extended operation under 10 CFR Part 54 should be eliminated. Accordingly, while the Commission has not assigned a single level of significance for the impacts of spent fuel and high-level waste disposal, this issue is considered in Category 1.

Table B-1. (cont.)

(c) The 100-year environmental dose commitment to the U.S. population from the fuel cycle, high-level waste and spent fuel disposal excepted, is calculated to be about 14,800 person rem, or 12 cancer fatalities, for each additional 20-year power reactor operating term. Much of this, especially the contribution of radon releases from mines and tailing piles, consists of tiny doses summed over large populations. This same dose calculation can theoretically be extended to include many tiny doses over additional thousands of years as well as doses outside the United States. The result of such a calculation would be thousands of cancer fatalities from the fuel cycle, but this result assumes that even tiny doses have some statistical adverse health effect that will never be mitigated (e.g., no cancer cure in the next thousand years), and that these doses projected over thousands of years are meaningful. However, these assumptions are questionable. In particular, science cannot rule out the possibility that there will be no cancer fatalities from these tiny doses. For perspective, the doses are very small fractions of regulatory limits, and even smaller fractions of natural background exposure to the same populations.

Nevertheless, despite all the uncertainty, some judgment as to the regulatory NEPA implications of these matters should be made, and it makes no sense to repeat the same judgment in every case. Even taking the uncertainties into account, the Commission concludes that these impacts are acceptable in that these impacts would not be sufficiently large to require the NEPA conclusion, for any plant, that the option of extended operation under 10 CFR Part 54 should be eliminated. Accordingly, while the Commission has not assigned a single level of significance for the collective effects of the fuel cycle, this issue is considered Category 1.

Appendix C

General Characteristics and Environmental Settings of Domestic Nuclear Power Plants

Appendix C

General Characteristics and Environmental Settings of Domestic Nuclear Power Plants

This appendix contains brief descriptions of each commercial nuclear power plant site in the United States. The material is intended to serve as an overview of the important characteristics of each plant and its environmental setting. The information was taken from the 1996 GEIS (NRC 1996) and updated with information available from recently published supplemental environmental impacts statements (SEISs), environmental assessments, CEC (2006), DOE/EIA (2007a,b), USCB (2007), EPA (2007), NRC (2008a,b, 2010), USFWS (2007), and USGS (2003).

ARKANSAS NUCLEAR ONE

Location:	Pope County, Arkansas 6 mi (10 km) WNW of Russellville Latitude 35.3100°N; longitude 93.2308°W
Licensee:	Entergy Nuclear Operations, Inc.

Unit Information	Unit 1	Unit 2
Docket Number:	50-313	50-368
Construction Permit:	1968	1972
Operating License:	1974	1978
Commercial Operation:	1974	1980
License Expiration:	2034	2038
Licensed Thermal Power (MWt):	2,568	3,026
Net Capacity (MWe):	843	995
Type of Reactor:	PWR	PWR
Nuclear Steam Supply System Vendor:	B&E	CE

Cooling Water System

Type: Unit 1: Once-through; Unit 2: Natural draft cooling tower
Source: Dardanelle Reservoir
Source Temperature Range: 40–83°F (4–28°C)
Condenser Flow Rate: 762,400 gpm (48.1 m^3/s) for Unit 1
422,000 gpm (26.6 m^3/s) for Unit 2
Design Condenser Temperature Rise: 15°F (8.3°C) for Unit 1
30.7°F (17.1°C) for Unit 2
Intake Structure: 4,400-ft (1,340-m) canal
Discharge Structure: 520-ft (158-m) canal

Site Information

Total Area: 1,164 ac (471 ha)
Exclusion Distance: 0.7 mi (1 km) radius
Low Population Zone: 4 mi (6.44 km) radius
Nearest City: Little Rock: 2000 population: 183,133
Site Topography: Flat
Surrounding Area Topography: Hilly to mountainous
Dominant Land Cover Within 5 mi (8 km): Forest, agriculture, open water
Level 1 Ecoregion within 5 mi (8 km): Eastern Temperate Forest

Level 3 Ecoregion within 5 mi (8 km): Arkansas Valley
Percent Wetland within 5 mi (8 km): 11.7, mostly lake
Nearby Features: The nearest town is London 2 mi (3 km) NW. The size of Lake Dardanelle is 37,000 ac (15,000 ha). The reservoir is part of the Arkansas River. The Missouri Pacific Railroad and U.S. Highway I-40 are just north of the site.
Population within an 50-mi (80-km) Radius: 267,664

BEAVER VALLEY POWER STATION

Location: Beaver County, Pennsylvania
25 mi (40 km) NW of Pittsburgh
Latitude 40.6219°N; longitude 80.4339°W
Licensee: FirstEnergy Nuclear Operating Company

Unit Information	Unit 1	Unit 2
Docket Number:	50-334	50-412
Construction Permit:	1970	1974
Operating License:	1976	1987
Commercial Operation:	1976	1987
License Expiration:	2036	2047
Licensed Thermal Power (MWt):	2,900	2,900
Net Capacity (MWe):	892	846
Type of Reactor:	PWR	PWR
Nuclear Steam Supply System Vendor:	WEST	WEST

Cooling Water System

Type: Natural draft cooling towers
Source: Ohio River
Source Temperature Range: 36.5–79.5°F (2.5–26.4°C)
Condenser Flow Rate: 480,400 gpm (30.31 m^3/s) each unit
Design Condenser Temperature Rise: 26°F (14°C)
Intake Structure: Concrete structure at river edge
Discharge Structure: At river edge

Site Information

Total Area: 453 ac (183 ha)
Exclusion Distance: 0.38 mi (0.61 km)
Low Population Zone: 3.60 mi (5.79 km)
Nearest City: Pittsburgh; 2000 population: 334,563
Site Topography: Flat
Surrounding Area Topography: Hilly
Dominant Land Cover within 5 mi (8 km): Forest, agriculture, developed: open space
Level 1 Ecoregion within 5 mi (8 km): Eastern Temperate Forest
Level 3 Ecoregion within 5 mi (8 km): Western Allegheny Plateau
Percent Wetland within 5 mi (8 km): 5.5, mostly riverine

Nearby Features: The nearest town is Midland 1 mi (1.6 km) NW. A large industrial area is about 1 mi (1.6 km) WNW. The Penn Central Railroad State Parks are within 10 mi (16 km).

Population within a 50 mi (80 km) Radius: 3,274,451

BRAIDWOOD STATION

Location:	Will County, Illinois 39 km (24 mi) SSW of Joliet Latitude 41.2436°N; longitude 88.2297°W
Licensee:	Exelon Generation Company

Unit Information	Unit 1	Unit 2
Docket Number:	50-456	50-457
Construction Permit:	1975	1975
Operating License:	1987	1988
Commercial Operation:	1988	1988
License Expiration:	2026	2027
Licensed Thermal Power (MWt):	3,587	3,587
Net Capacity (MWe):	1,178	1,152
Type of Reactor:	PWR	PWR
Nuclear Steam Supply System Vendor:	WEST	WEST

Cooling Water System

Type: Closed-cycle cooling pond
Source: Kankakee River
Source Temperature Range: 32–87°F (0–31°C)
Condenser Flow Rate: 729,800 gpm (46.05 m^3/s)
Design Condenser Temperature Rise: 21°F (12°C)
Intake Structure: Concrete structure at lake shore
Discharge Structure: Surface discharge flume to lake

Site Information

Total Area: 4,457 ac (1,804 ha)
Exclusion Distance: 0.3 mi (0.48 km) minimum
Low Population Zone: 1.125 mi (1.810 km) radius
Nearest City: Joliet; 2000 population: 106,221
Site Topography: Flat to rolling
Surrounding Area Topography: Flat to rolling
Dominant Land Cover within 5 mi (8 km): Agriculture, forest, developed: high, medium, low density
Level 1 Ecoregion within 5 mi (8 km): Eastern Temperate Forest
Level 3 Ecoregion within 5 mi (8 km): Central Corn Belt Plains

Percent Wetland within 5 mi (8 km): 11.4, mostly lake
Nearby Features: The nearest town is Godley 0.5 mi (0.8 km) SW. There are 4 State parks within 10 mi (16 km). Midewin National Tallgrass Prairie and Abraham Lincoln National Cemetery are about 8 mi (13 km) NE. Dresden Nuclear Power Station is about 10 mi (16 km) N, and LaSalle County Station (nuclear) is about 20 mi (32 km) WSW. The Illinois Central Gulf Railroad is just NW. U.S. Highway I-55 is about 2 mi (3 km) NW.
Population within an 80 km (50 mi) Radius: 4,272,003

BROWNS FERRY NUCLEAR PLANT

Location: Limestone County, Alabama
16 km (10 mi) NW of Decatur
Latitude 34.7042°N; longitude 87.1186°W
Licensee: Tennessee Valley Authority

Unit Information	Unit 1	Unit 2	Unit 3
Docket Number:	50-259	50-260	50-296
Construction Permit:	1967	1967	1968
Operating License:	1973	1974	1976
Commercial Operation:	1974	1975	1977
License Expiration:	2033	2034	2036
Licensed Thermal Power (MWt):	3,458	3,458	3,458
Net Capacity (MWe):	1,065	1,104	1,115
Type of Reactor:	BWR	BWR	BWR
Nuclear Steam Supply System Vendor:	GE	GE	GE

Cooling Water System

Type: Once-through with helper towers
Source: Tennessee River
Source Temperature Range: 40–90°F (4–32°C)
Condenser Flow Rate: 734,000 gpm (139 m^3/s); for all three units
Design Condenser Temperature Rise: 28.7°F (15.9°C)
Intake Structure: Concrete structure in small inlet
Discharge Structure: Diffuser pipes

Site Information

Total Area: 840 ac (340 ha)
Exclusion Distance: 0.76 mi (1.22 km) radius
Low Population Zone: 7 mi (11.3 km)
Nearest City: Huntsville; 2000 population: 158,216
Site Topography: Flat
Surrounding Area Topography: Flat to rolling
Dominant Land Cover within 5 mi (8 km): Agriculture, open water, forest
Level 1 Ecoregion within 5 mi (8 km): Eastern Temperate Forest
Level 3 Ecoregion within 5 mi (8 km): Interior Plateau

Percent Wetland within 5 mi (8 km): 42.2, mostly lake (some freshwater forested/shrub wetland)

Nearby Features: The nearest town is Lawngate 1 mi (1.6 km) NE. The Redstone Arsenal is 25 mi (40 km) E. The Southern Railroad is 6 mi (10 km) S, and the Louisville and Nashville Railroad is 6 mi (10 km) E. Two wildlife management areas are located within 3 mi (5 km) of the plant.

Population within a 50 mi (80 km) Radius: 872,478

BRUNSWICK STEAM ELECTRIC PLANT

Location: Brunswick County, North Carolina
16 mi (26 km) S of Wilmington
Latitude 33.9583°N; longitude 78.0106°W
Licensee: Progress Energy

Unit Information	Unit 1	Unit 2
Docket Number:	50-325	50-324
Construction Permit:	1967	1968
Operating License:	1976	1974
Commercial Operation:	1977	1975
License Expiration:	2036	2034
Licensed Thermal Power (MWt):	2,923	2,923
Net Capacity (MWe):	938	937
Type of Reactor:	BWR	BWR
Nuclear Steam Supply System Vendor:	GE	GE

Cooling Water System

Type: Once-through
Source: Cape Fear River
Source Temperature Range: 40–86°F (4–30°C)
Condenser Flow Rate: 675,000 gpm (42.6 m^3/s)
Design Condenser Temperature Rise: 17°F (9°C)
Intake Structure: 3 mi (5 km) canal from Cape Fear River
Discharge Structure: 6 mi (10 km) canal to Atlantic Ocean

Site Information

Total Area: 1,200 ac (490 ha)
Exclusion Distance: 0.57 mi (0.92 km)
Low Population Zone: 2 mi (3.22 km)
Nearest City: Wilmington; 2000 population: 75,838
Site Topography: Flat
Surrounding Area Topography: Flat
Dominant Land Cover within 5 mi (8 km): Wetland, open water, forest
Level 1 Ecoregion within 5 mi (8 km): Eastern Temperate Forest
Level 3 Ecoregion within 5 mi (8 km): Middle Atlantic Coastal Plain

Percent Wetland within 5 mi (8 km): 60.5, mostly estuarine and marine deepwater; freshwater forested/shrub wetland; estuarine and marine wetland
Nearby Features: The nearest town is Southport 3 mi (5 km) S. Sunny Point Military Ocean Terminal is about 5 mi (8 km) N.
Population within a 50 mi (80 km) Radius: 361,872

BYRON STATION

Location:	Ogle County, Illinois 17 mi (27 km) SW of Rockford Latitude 42.0750°N; longitude 89.2811°W
Licensee:	Exelon Generation Company

Unit Information	Unit 1	Unit 2
Docket Number:	50-454	50-455
Construction Permit:	1975	1975
Operating License:	1985	1987
Commercial Operation:	1985	1987
License Expiration:	2025	2027
Licensed Thermal Power (MWt):	3,587	3,587
Net Capacity (MWe):	1,164	1,136
Type of Reactor:	PWR	PWR
Nuclear Steam Supply System Vendor:	WEST	WEST

Cooling Water System

Type: Natural draft towers
Source: Rock River
Source Temperature Range: Not available
Condenser Flow Rate: 632,000 gpm (39.9 m^3/s)
Design Condenser Temperature Rise: 24°F (13°C)
Intake Structure: Concrete structure on river bank
Discharge Structure: Discharged to river

Site Information

Total Area: 1,398 ac (565.8 ha)
Exclusion Distance: 0.26 mi (0.42 km)
Low Population Zone: 3 mi (4.83 km)
Nearest City: Rockford; 2000 population: 150,115
Site Topography: Rolling
Surrounding Area Topography: Rolling
Dominant Land Cover within 5 mi (8 km): Agriculture, forest, developed: open space
Level 1 Ecoregion within 5 mi (8 km): Eastern Temperate Forest
Level 3 Ecoregion within 5 mi (8 km): Central Corn Belt Plains
Percent Wetland within 5 mi (8 km): 3.6, mostly lake

Nearby Features: The nearest town is Byron about 3 mi (5 km) NNE. The Chicago Milwaukee and the St. Paul and Pacific Railroads are about 4 mi (6 km) NNE. White Pines State Park is about 11 mi (18 km) WSW.

Population within a 50 mi (80 km) Radius: 1,300,282

CALLAWAY PLANT

Location: Callaway County, Missouri
10 mi (16 km) SE of Fulton
Latitude 38.7622°N; longitude 91.7817°W
Licensee: Ameren Corporation

Unit Information	Unit 1
Docket Number:	50-483
Construction Permit:	1976
Operating License:	1984
Commercial Operation:	1984
License Expiration:	2024
Licensed Thermal Power (MWt):	3,565
Net Capacity (MWe):	1,190
Type of Reactor:	PWR
Nuclear Steam Supply System Vendor:	WEST

Cooling Water System

Type: Natural draft cooling tower
Source: Missouri River
Source Temperature Range: Not available
Condenser Flow Rate: 530,000 gpm (33 m^3/s)
Design Condenser Temperature Rise: 30°F (17°C)
Intake Structure: Intake from river
Discharge Structure: Discharged to river

Site Information

Total Area: 5,228 ac (2,115.8 ha)
Exclusion Distance: 0.75 mi (1.21 km) radius
Low Population Zone: 2.50 mi (4.02 ha)
Nearest City: Columbia; 2000 population: 84,531
Site Topography: Flat, on a small plateau
Surrounding Area Topography: Rolling to hilly
Dominant Land Cover within 5 mi (8 km): Forest, agriculture, developed: open space
Level 1 Ecoregion within 5 mi (8 km): Eastern Temperate Forest
Level 3 Ecoregion within 5 mi (8 km): Interior River Valley and Hills
Percent Wetland within 5 mi (8 km): 4.5, mostly freshwater forested/shrub wetland; riverine

Nearby Features: The nearest town is Portland 5 mi (8 km) SE. The Missouri, Kansas, and Texas Railroad is about 3 mi (5 km) S, and the Missouri Pacific Railroad is about 6 mi (10 km) S. U.S. Highway I-70 is about 10 mi (16 km) N.

Population within a 50 mi (80 km) Radius: 491,072

CALVERT CLIFFS NUCLEAR POWER PLANT

Location: Calvert County, Maryland
35 mi (56 km) S of Annapolis
Latitude 38.4347°N; longitude 76.4419°W
Licensee: Constellation Energy

Unit Information	Unit 1	Unit 2
Docket Number:	50-317	50-318
Construction Permit:	1969	1969
Operating License:	1974	1976
Commercial Operation:	1975	1977
License Expiration:	2034	2036
Licensed Thermal Power (MWt):	2,700	2,700
Net Capacity (MWe):	873	862
Type of Reactor:	PWR	PWR
Nuclear Steam Supply System Vendor:	CE	CE

Cooling Water System

Type: Once-through
Source: Chesapeake Bay
Source Temperature Range: 34–87°F (1–31°C)
Condenser Flow Rate: 1,200,000 gpm (76 m^3/s) each unit
Design Condenser Temperature Rise: 12°F (6.7°C).
Intake Structure: 4,500 ft (1,372 m) from shore
Discharge Structure: 850 ft (260 m) from shore

Site Information

Total Area: 2,108 ac (853 ha)
Exclusion Distance: 0.67 mi (1.08 km) radius
Low Population Zone: 2 mi (3.2 km)
Nearest City: Washington, D.C.; 2000 population: 572,059
Site Topography: Rolling
Surrounding Area Topography: Rolling
Dominant Land Cover within 5 mi (8 km): Open water, forest, agriculture
Level 1 Ecoregion within 5 mi (8 km): Eastern Temperate Forest
Level 3 Ecoregion within 5 mi (8 km): Southeastern Plains; Middle Atlantic Coastal Plain
Percent Wetland within 5 mi (8 km): 66, mostly estuarine and marine deepwater

Nearby Features: The nearest town is Long Beach 1 mi (1.6 km) NNW. Calvert Cliffs State Park is about 4 mi (6 km) SSE. A naval ordinance facility is 7 mi (11 km) SSW.

Population within a 50 mi (80 km) Radius: 3,919,397

CATAWBA NUCLEAR STATION

Location:	York County, South Carolina 6 mi (10 km) NNW of Rock Hill Latitude 35.0514°N; longitude 81.0708°W
Licensee:	Duke Energy Power Corporation

Unit Information	Unit 1	Unit 2
Docket Number:	50-413	50-414
Construction Permit:	1975	1975
Operating License:	1985	1986
Commercial Operation:	1985	1986
License Expiration:	2043	2043
Licensed Thermal Power (MWt):	3,411	3,411
Net Capacity (MWe):	1,129	1,129
Type of Reactor:	PWR	PWR
Nuclear Steam Supply System Vendor:	WEST	WEST

Cooling Water System

Type: Mechanical draft towers
Source: Lake Wylie
Source Temperature Range: 43–83°F (6–28°C)
Condenser Flow Rate: 660,000 gpm (42 m^3/s) each unit
Design Condenser Temperature Rise: 24°F (13°C)
Intake Structure: Skimmer wall on cove of the lake
Discharge Structure: On another cove of the lake

Site Information

Total Area: 391 ac (158 ha)
Exclusion Distance: 2,500 ft (0.76 km; 0.47 mi) radius
Low Population Zone: 3.8 mi (6.12 km) radius
Nearest City: Charlotte, North Carolina; 2000 population: 540,828
Site Topography: Rolling
Surrounding Area Topography: Rolling
Dominant Land Cover within 5 mi (8 km): Forest, agriculture, developed: open space
Level 1 Ecoregion within 5 mi (8 km): Eastern Temperate Forest
Level 3 Ecoregion within 5 mi (8 km): Piedmont
Percent Wetland within 5 mi (8 km): 12.9, mostly lake

Nearby Features: The nearest town is Rock Hill 6 mi (10 km) SSE. U.S. Highway I-77 is about 6 mi (10 km) E and I-85 is about 17 mi (27 km) N. The Southern Railway is 5 mi (8 km) S.

Population within a 50 mi (80 km) Radius: 2,041,465

CLINTON POWER STATION

Location: DeWitt County, Illinois
6 mi (10 km) E of Clinton
Latitude 40.1731°N; longitude 88.8342°W
Licensee: Exelon Generation Company

Unit Information	Unit 1
Docket Number:	50-461
Construction Permit:	1976
Operating License:	1987
Commercial Operation:	1987
License Expiration:	2026
Licensed Thermal Power (MWt):	3,473
Net Capacity (MWe):	1,065
Type of Reactor:	BWR
Nuclear Steam Supply System Vendor:	GE

Cooling Water System

Type: Once-through
Source: Salt Creek
Source Temperature Range: 32–83°F (0–28°C)
Condenser Flow Rate: 568,701 gpm (35.89 m^3/s)
Design Condenser Temperature Rise: 23°F (13°C)
Intake Structure: Concrete structure at shoreline of North Fork Salt Creek
Discharge Structure: 3-mi (5-km) flume discharging to Salt Creek

Site Information

Total Area: 14,090 ac (5,702 ha)
Exclusion Distance: 0.60 mi (0.97 km) radius
Low Population Zone: 2.5 mi (4.02 km) radius
Nearest City: Decatur; 2000 population: 81,860
Site Topography: Flat
Surrounding Area Topography: Flat
Dominant Land Cover within 5 mi (8 km): Agriculture, forest, open water
Level 1 Ecoregion within 5 mi (8 km): Eastern Temperate Forest
Level 3 Ecoregion within 5 mi (8 km): Central Corn Belt Plains
Percent Wetland within 5 mi (8 km): 9, mostly lake

Nearby Features: The nearest town is DeWitt 2 mi (3 km) ENE. Weldon Springs State Park is 6 mi (10 km) SW. The Illinois Central Gulf Railroad crosses the site. U.S. highway I-74 is 11 mi (18 km) NE. A dam on Salt Creek near the site creates the reservoir Lake Clinton for the cooling water system.

Population within a 50 mi (80 km) Radius: 789,754

COLUMBIA GENERATING STATION

Location:	Benton County, Washington 12 mi (19 km) NW of Richland Latitude 46.4714°N; longitude 119.3331°W
Licensee:	Energy Northwest

Unit Information	Unit 2
Docket Number:	50-397
Construction Permit:	1973
Operating License:	1984
Commercial Operation:	1984
License Expiration:	2023
Licensed Thermal Power (MWt):	3,323
Net Capacity (MWe):	1,131
Type of Reactor:	BWR
Nuclear Steam Supply System Vendor:	GE

Cooling Water System

Type: Mechanical draft cooling towers
Source: Columbia River
Source Temperature Range: 38–64°F (3–18°C)
Condenser Flow Rate: 550,000 gpm (35 m^3/s)
Design Condenser Temperature Rise: 28.7°F (15.9°C)
Intake Structure: 2 perforated pipe inlets supported offshore above the river bed 900 ft (270 m) from pump structure on river bank
Discharge Structure: Buried 3 mi (5 km) pipeline, terminating at the river bed 175 ft (53 m) from the shoreline

Site Information

Total Area: 1,089 ac (441 ha)
Exclusion Distance: 1.21 mi (1.95 km) radius
Low Population Zone: 3 mi (4.83 km)
Nearest City: Spokane; 2000 population: 195,629
Site Topography: Flat
Surrounding Area Topography: Flat
Dominant Land Cover within 5 mi (8 km): Shrub/scrub, open water, agriculture
Level 1 Ecoregion within 5 mi (8 km): North American Desert

Level 3 Ecoregion within 5 mi (8 km): Columbia Plateau
Percent Wetland within 5 mi (8 km): 5.6, mostly lake
Nearby Features: The nearest town is Richland 9 mi (14 km) S. The site is in the SE part of the Hanford Reservation.
Population within a 50 mi (80 km) Radius: 360,573

COMANCHE PEAK STEAM ELECTRIC STATION

Location: Somervell County, Texas
40 mi (64 km) SW of Fort Worth
Latitude 32.2983°N; longitude 97.7856°W

Licensee: Luminant Energy Co.

Unit Information	Unit 1	Unit 2
Docket Number:	50-445	50-446
Construction Permit:	1974	1974
Operating License:	1990	1993
Commercial Operation:	1990	1993
License Expiration:	2030	2033
Licensed Thermal Power (MWt):	3,458	3,458
Net Capacity (MWe):	1,200	1,150
Type of Reactor:	PWR	PWR
Nuclear Steam Supply System Vendor:	WEST	WEST

Cooling Water System

Type: Once-through
Source: Squaw Creek Reservoir
Source Temperature Range: Not available
Condenser Flow Rate: 1,030,000 gpm (65 m^3/s)
Design Condenser Temperature Rise: 15°F (8°C)
Intake Structure: On shore of reservoir
Discharge Structure: Canal to reservoir

Site Information

Total Area: 7,669 ac (3,104 ha)
Exclusion Distance: 0.96 mi (1.54 km) minimum
Low Population Zone: 4 mi (6.44 km) radius
Nearest City: Fort Worth; 2000 population: 534,694
Site Topography: Flat, with hills rising from the reservoir
Surrounding Area Topography: Rolling to hilly
Dominant Land Cover within 5 mi (8 km): Herbaceous, forest, open water
Level 1 Ecoregion within 5 mi (8 km): Great Plains
Level 3 Ecoregion within 5 mi (8 km): Cross Timbers
Percent Wetland within 5 mi (8 km): 8.8, mostly lake

Nearby Features: The nearest town is Glen Rose 5 mi (8 km) SSE. Dinosaur Valley State Park is 5 mi (8 km) SW. A 26-in. (66-cm) oil pipeline is very near the site, and a 36-in. (91-cm) natural gas line is about 2 mi (3 km) from the site.

Population within a 50 mi (80 km) Radius: 1,431,094

COOPER NUCLEAR STATION

Location:	Nemaha County, Nebraska 23 mi (37 km) S of Nebraska City Latitude 40.3619°N; longitude 95.6411°W
Licensee:	Nebraska Public Power District

Unit Information	Unit 1
Docket Number:	50-298
Construction Permit:	1968
Operating License:	1974
Commercial Operation:	1974
License Expiration:	2034
Licensed Thermal Power (MWt):	2,419
Net Capacity (MWe):	830
Type of Reactor:	BWR
Nuclear Steam Supply System Vendor:	GE

Cooling Water System

Type: Once-through
Source: Missouri River
Source Temperature Range: 34–73°F (1–23°C)
Condenser Flow Rate: 631,000 gpm (39.8 m^3/s)
Design Condenser Temperature Rise: 18°F (10°C)
Intake Structure: At shoreline
Discharge Structure: At shoreline

Site Information

Total Area: 1,090 ac (441 ha)
Exclusion Distance: 0.68 mi (1.09 km)
Low Population Zone: 1 mi (1.61 km) radius
Nearest City: Lincoln; 2000 population: 225,581
Site Topography: Flat
Surrounding Area Topography: Flat
Dominant Land Cover within 5 mi (8 km): Agriculture, wetland, forest
Level 1 Ecoregion within 5 mi (8 km): Great Plains
Level 3 Ecoregion within 5 mi (8 km): Western Corn Belt Plains
Percent Wetland within 5 mi (8 km): 6.8, mostly freshwater forested/shrub wetland; riverine

Nearby Features: The nearest town is Nemaha about 1 mi (1.6 km) S. A railroad runs just W of the site. Indian Cave State Park is about 8 mi (13 km) SSE.

Population within a 50 mi (80 km) Radius: 156,157

CRYSTAL RIVER NUCLEAR POWER PLANT

Location: Citrus County, Florida
7 mi (11 km) NW of Crystal River
Latitude 28.9572°N; longitude 82.6989°W

Licensee: Progress Energy

Unit Information	Unit 3
Docket Number:	50-302
Construction Permit:	1968
Operating License:	1977
Commercial Operation:	1977
License Expiration:	2016
Licensed Thermal Power (MWt):	2,609
Net Capacity (MWe):	838
Type of Reactor:	PWR
Nuclear Steam Supply System Vendor:	B&W

Cooling Water System

Type: Once-through
Source: Gulf of Mexico
Source Temperature Range: 87°F (31°C) maximum
Condenser Flow Rate: 680,000 gpm (43 m^3/s)
Design Condenser Temperature Rise: 17.1°F (9.5°C)
Intake Structure: 16,000 ft (4,900 m) from shoreline
Discharge Structure: 13,000 ft (4,000 m) canal

Site Information

Total Area: 4,738 ac (1,917 ha)
Exclusion Distance: 0.83 mi (1.34 km) radius
Low Population Zone: 5 mi (8.05 km)
Nearest City: Gainesville; 2000 population: 95,447
Site Topography: Swamps and marshland
Surrounding Area Topography: Flat
Dominant Land Cover within 5 mi (8 km): Open water, wetland, forest
Level 1 Ecoregion within 5 mi (8 km): Eastern Temperate Forest
Level 3 Ecoregion within 5 mi (8 km): Southern Coastal Plain
Percent Wetland within 5 mi (8 km): 65.2, mostly estuarine and marine deepwater

Nearby Features: The nearest town is Crystal River about 7 mi (11 km) SE. Units 1 and 2 are coal-fired plants and share a common intake and discharge with the nuclear unit.

Population within a 50 mi (80 km) Radius: 1,273,146

DAVIS-BESSE NUCLEAR POWER STATION

Location:	Ottawa County, Ohio 21 mi (34 km) E of Toledo Latitude 41.5972°N; longitude 83.0864°W
Licensee:	FirstEnergy Nuclear Operating Co.

Unit Information	Unit 1
Docket Number:	50-346
Construction Permit:	1971
Operating License:	1977
Commercial Operation:	1978
License Expiration:	2017
Licensed Thermal Power (MWt):	2,817
Net Capacity (MWe):	893
Type of Reactor:	PWR
Nuclear Steam Supply System Vendor:	B&W

Cooling Water System

Type: Natural draft cooling tower
Source: Lake Erie
Source Temperature Range: 34–73°F (1–23°C)
Condenser Flow Rate: 480,000 gpm (30 m^3/s)
Design Condenser Temperature Rise: 26°F (14°C)
Intake Structure: Submerged intake about 3,000 ft (900 m) offshore
Discharge Structure: Submerged discharge about 930 ft (280 m) offshore

Site Information

Total Area: 954 ac (386 ha)
Exclusion Distance: 0.45 mi (0.72 km) radius
Low Population Zone: 2 mi (3.22 km)
Nearest City: Toledo; 2000 population: 313,619
Site Topography: Flat
Surrounding Area Topography: Flat
Dominant Land Cover within 5 mi (8 km): Open water, agriculture, wetland
Level 1 Ecoregion within 5 mi (8 km): Eastern Temperate Forest
Level 3 Ecoregion within 5 mi (8 km): Huron/Erie Lake Plains
Percent Wetland within 5 mi (8 km): 66.6, mostly lake

Nearby Features: The nearest town is Oak Harbor about 6 mi (10 km) SW. Several wildlife refuge areas are within 5 mi (8 km) of the site.

Population within a 50 mi (80 km) Radius: 2,617,550

DIABLO CANYON POWER PLANT

Location: San Luis Obispo County, California
12 mi (19 km) W of San Luis Obispo
Latitude 35.2117°N; longitude 120.8544°W
Licensee: Pacific Gas and Electric Co.

Unit Information	Unit 1	Unit 2
Docket Number:	50-275	50-323
Construction Permit:	1968	1970
Operating License:	1984	1985
Commercial Operation:	1985	1986
License Expiration:	2024	2025
Licensed Thermal Power (MWt):	3,411	3,411
Net Capacity (MWe):	1,122	1,118
Type of Reactor:	PWR	PWR
Nuclear Steam Supply System Vendor:	WEST	WEST

Cooling Water System

Type: Once-through
Source: Pacific Ocean
Source Temperature Range: 50–63°F (10–17°C)
Condenser Flow Rate: 863,000 gpm (54.5 m^3/s)
Design Condenser Temperature Rise: 18°F (10°C)
Intake Structure: Reinforced-concrete structure located at shoreline in a cove with artificial breakwater wall
Discharge Structure: Reinforced-concrete structure drops water in stair-step type weir overflow from elevation 70 ft (21 m) to the ocean and discharges on the surface at the shoreline

Site Information

Total Area: 750 ac (300 ha)
Exclusion Distance: 0.50 mi (0.80 km)
Low Population Zone: 6 mi (9.66 km)
Nearest City: Santa Barbara; 2000 population: 92,325
Site Topography: Hilly
Surrounding Area Topography: Hilly to mountainous
Dominant Land Cover within 5 mi (8 km): Open water, forest, shrub/scrub

Level 1 Ecoregion within 5 mi (8 km): Mediterranean California
Level 3 Ecoregion within 5 mi (8 km): Southern and Central California Chaparral and Oak Woodlands
Percent Wetland within 5 mi (8 km): 54.6, mostly estuarine and marine deepwater
Nearby Features: Site is remote, the nearest town being San Obispo 12 mi (19 km) E. Beaches 7–15 mi (11–24 km) ESE have an influx of summer visitors. Pismo Beach State Park and Morro Bay State Park are within 15 mi (24 km). Vandenberg Air Base is 35 mi (56 km) ESE.
Population within a 50 mi (80 km) Radius: 836,031

DONALD C. COOK NUCLEAR PLANT

Location: Berrien County, Michigan
10 mi (16 km) S of St. Joseph
Latitude 41.9761°N; longitude 86.5664°W
Licensee: Indiana Michigan Power Co.

Unit Information	Unit 1	Unit 2
Docket Number:	50-315	50-316
Construction Permit:	1969	1969
Operating License:	1974	1977
Commercial Operation:	1975	1978
License Expiration:	2034	2037
Licensed Thermal Power (MWt):	3,304	3,468
Net Capacity (MWe):	1,009	1,060
Type of Reactor:	PWR	PWR
Nuclear Steam Supply System Vendor:	WEST	WEST

Cooling Water System

Type: Once-through
Source: Lake Michigan
Source Temperature Range: 34–73°F (1–23°C)
Condenser Flow Rate: 1.6 million gal/min (both units)
Design Condenser Temperature Rise: 20°F (11°C)
Intake Structure: Intake cribs 2,250 ft (686 m) from shore
Discharge Structure: 1,150 ft (351 m) from shore

Site Information

Total Area: 650 ac (260 ha)
Exclusion Distance: 0.38 mi (0.61 km)
Low Population Zone: 2 mi (3.22 km)
Nearest City: South Bend, Indiana; 2000 population: 107,789
Site Topography: Rolling
Surrounding Area Topography: Flat to rolling
Dominant Land Cover within 5 mi (8 km): Open water, agriculture, forest
Level 1 Ecoregion within 5 mi (8 km): Eastern Temperate Forest
Level 3 Ecoregion within 5 mi (8 km): S. Michigan/N. Indiana Drift Plains
Percent Wetland within 5 mi (8 km): 53.6, mostly lake

Nearby Features: The nearest town is Livingston 1 mi (1.6 km) SW. The Chesapeake and Ohio Railroad and U.S. Highway I-94 are just E of the site. Warren Dunes State Park is about 5 mi (8 km) SSW.

Population within a 50 mi (80 km) Radius: 1,447,303

DRESDEN NUCLEAR POWER STATION

Location:	Grundy County, Illinois 9 mi (14 km) E of Morris Latitude 41.3897°N; longitude 88.2711°W
Licensee:	Exelon Generation Company

Unit Information	Unit 2	Unit 3
Docket Number:	50-237	50-249
Construction Permit:	1966	1966
Operating License:	1969	1971
Commercial Operation:	1970	1971
License Expiration:	2029	2031
Licensed Thermal Power (MWt):	2,957	2,957
Net Capacity (MWe):	867	867
Type of Reactor:	BWR	BWR
Nuclear Steam Supply System Vendor:	GE	GE

Cooling Water System

Type: Cooling lake and spray canal; mechanical draft towers
Source: Kankakee River
Source Temperature Range: 40–85°F (4–29°C)
Condenser Flow Rate: 940,000 gpm (both units)
Design Condenser Temperature Rise: Not available
Intake Structure: Canal from Kankakee River to a crib house
Discharge Structure: A canal carries water to a cooling lake of about 1,275 ac (516 ha)

Site Information

Total Area: 2,500 ac (1,012 ha)
Exclusion Distance: 0.5 mi (0.8 km) radius
Low Population Zone: 5 mi (8 km)
Nearest City: Joliet; 2000 population: 106,221
Site Topography: Flat
Surrounding Area Topography: Rolling
Dominant Land Cover within 5 mi (8 km): Agriculture, herbaceous, forest
Level 1 Ecoregion within 5 mi (8 km): Eastern Temperate Forest
Level 3 Ecoregion within 5 mi (8 km): Central Corn Belt Plains
Percent Wetland within 5 mi (8 km): 22, mostly lake

Nearby Features: The nearest town is Channahon 3 mi (5 km) NNE. Braidwood Station (nuclear plant) is about 10 mi (16 km) S and LaSalle County Station (nuclear plant) is about 22 mi (35 km) SW.

Population within a 50 mi (80 km) Radius: 7,337,564

DUANE ARNOLD ENERGY CENTER

Location:	Linn County, Iowa 8 mi (13 km) NW of Cedar Rapids Latitude 42.1006°N; longitude 91.7772°W
Licensee:	Florida Power & Light Co.

Unit Information	Unit 1
Docket Number:	50-331
Construction Permit:	1970
Operating License:	1974
Commercial Operation:	1975
License Expiration:	2034
Licensed Thermal Power (MWt):	1,912
Net Capacity (MWe):	640
Type of Reactor:	BWR
Nuclear Steam Supply System Vendor:	GE

Cooling Water System

Type: Mechanical draft cooling towers
Source: Cedar River
Source Temperature Range: 32–89°F (0–32°C)
Condenser Flow Rate: 290,000 gpm (18 m^3/s)
Design Condenser Temperature Rise: 25°F (14°C)
Intake Structure: Structure on river shoreline
Discharge Structure: Canal to shoreline

Site Information

Total Area: 500 ac (200 ha)
Exclusion Distance: 0.27 mi (0.43 km)
Low Population Zone: 6 mi (9.66 km)
Nearest City: Cedar Rapids; 2000 population: 120,758
Site Topography: Flat
Surrounding Area Topography: Rolling to hilly
Dominant Land Cover within 5 mi (8 km): Agriculture, forest, wetland
Level 1 Ecoregion within 5 mi (8 km): Great Plains
Level 3 Ecoregion within 5 mi (8 km): Western Corn Belt Plains
Percent Wetland within 5 mi (8 km): 11.7, mostly freshwater forested/shrub wetland

Nearby Features: The nearest town is Palo about 2 mi (3 km) SW. Several wildlife refuge areas are within 10 mi (16 km) of the site.

Population within a 50 mi (80 km) Radius: 613,736

EDWIN I. HATCH NUCLEAR PLANT

Location: Appling County, Georgia
11 mi (18 km) N of Baxley
Latitude 31.9342°N; longitude 82.3444°W
Licensee: Southern Nuclear Operating Company

Unit Information	Unit 1	Unit 2
Docket Number:	50-321	50-366
Construction Permit:	1969	1972
Operating License:	1974	1978
Commercial Operation:	1975	1979
License Expiration:	2034	2038
Licensed Thermal Power (MWt):	2,804	2,804
Net Capacity (MWe):	876	883
Type of Reactor:	BWR	BWR
Nuclear Steam Supply System Vendor:	GE	GE

Cooling Water System

Type: Mechanical draft towers
Source: Altamaha River
Source Temperature Range: 43–90°F (6–32°C)
Condenser Flow Rate: 556,000 gpm (35.1 m^3/s) each unit
Design Condenser Temperature Rise: 20°F (11°C)
Intake Structure: At edge of river
Discharge Structure: 120 ft (37 m) from shore

Site Information

Total Area: 2,244 ac (908 ha)
Exclusion Distance: 0.78 mi (1.26 km)
Low Population Zone: 0.78 mi (1.26 km)
Nearest City: Savannah; 2000 population: 131,510
Site Topography: Flat to rolling
Surrounding Area Topography: Flat to rolling
Dominant Land Cover within 5 mi (8 km): Forest, wetland, agriculture
Level 1 Ecoregion within 5 mi (8 km): Eastern Temperate Forest
Level 3 Ecoregion within 5 mi (8 km): Southeastern Plains; Southern Coastal Plain
Percent Wetland within 5 mi (8 km): 23.9, mostly freshwater forested/shrub wetland

Nearby Features: The nearest town is Cedar Crossing about 7 mi (11 km) NNW.
U.S. Highway 1 is just W of the site.
Population within a 50 mi (80 km) Radius: 366,508

ENRICO FERMI ATOMIC POWER PLANT

Location:	Monroe County, Michigan 30 mi (48 km) SW of Detroit Latitude 41.9631°N; longitude 83.2578°W
Licensee:	Detroit Edison Co.

Unit Information	Unit 2
Docket Number:	50-341
Construction Permit:	1972
Operating License:	1985
Commercial Operation:	1988
License Expiration:	2025
Licensed Thermal Power (MWt):	3,292
Net Capacity (MWe):	1,122
Type of Reactor:	BWR
Nuclear Steam Supply System Vendor:	GE

Cooling Water System

Type: Natural draft cooling towers
Source: Lake Erie
Source Temperature Range: 34–76°F (1–24°C)
Condenser Flow Rate: 836,000 gpm (52.80 m^3/s)
Design Condenser Temperature Rise: 18°F (10°C)
Intake Structure: At edge of lake
Discharge Structure: To the lake via a 50-ac (20-ha) pond

Site Information

Total Area: 1,120 ac (453 ha)
Exclusion Distance: 0.57 mi (0.92 km)
Low Population Zone: 3 mi (4.83 km)
Nearest City: Detroit; 2000 population: 951,270
Site Topography: Flat
Surrounding Area Topography: Flat to rolling
Dominant Land Cover within 5 mi (8 km): Open water, agriculture, developed: high, medium, low density
Level 1 Ecoregion within 5 mi (8 km): Eastern Temperate Forest
Level 3 Ecoregion within 5 mi (8 km): Huron/Erie Lake Plains

Percent Wetland within 5 mi (8 km): 57.9, mostly lake
Nearby Features: The town of Stony Point is adjacent to the site to the S. Sterling State Park and General Custer Historical Site are about 5 mi (8 km) SW.
Population within a 50 mi (80 km) Radius: 7,803,464

JAMES A. FITZPATRICK NUCLEAR POWER PLANT

Location:	Oswego County, New York 6 mi (10 km) NE of Oswego Latitude 43.5239°N; longitude 76.3983°W
Licensee:	Entergy Nuclear Operations, Inc.

Unit Information	Unit 1
Docket Number:	50-333
Construction Permit:	1970
Operating License:	1974
Commercial Operation:	1975
License Expiration:	2034
Licensed Thermal Power (MWt):	2,536
Net Capacity (MWe):	852
Type of Reactor:	BWR
Nuclear Steam Supply System Vendor:	GE

Cooling Water System

Type: Once-through
Source: Lake Ontario
Source Temperature Range: 32–68°F (0–20°C)
Condenser Flow Rate: 352,600 gpm (22.25 m^3/s)
Design Condenser Temperature Rise: 32°F (18°C)
Intake Structure: 900 ft (274 m) from shore
Discharge Structure: 1,400 ft (427 m) from shore

Site Information

Total Area: 702 ac (284 ha)
Exclusion Distance: 3,000 ft (914 m) to the east, over 1 mi (1.6 km) to the west, and about 1.5 mi (2.4 km) to the southern site boundary
Low Population Zone: 3.4 mi (5.47 km)
Nearest City: Syracuse; 2000 population: 147,306
Site Topography: Flat to rolling
Surrounding Area Topography: Rolling
Dominant Land Cover within 5 mi (8 km): Open water, forest, agriculture
Level 1 Ecoregion within 5 mi (8 km): Eastern Temperate Forest
Level 3 Ecoregion within 5 mi (8 km): Eastern Great Lakes and Hudson Lowlands

Percent Wetland within 5 mi (8 km): 65.4, mostly lake
Nearby Features: The nearest town is Lakeview about 1 mi (1.6 km) WSW. Fort Ontario is about 5 mi (8 km) SW. Nine Mile Point Nuclear Station is about 0.5 mi (0.8 km) W.
Population within a 50 mi (80 km) Radius: 914,668

JOSEPH M. FARLEY NUCLEAR PLANT

Location:	Houston County, Alabama 16 mi (26 km) E of Dothan Latitude 31.2228°N; longitude 85.1125°W
Licensee:	Southern Nuclear Operating Company

Unit Information	Unit 1	Unit 2
Docket Number:	50-348	50-364
Construction Permit:	1972	1972
Operating License:	1977	1981
Commercial Operation:	1977	1981
License Expiration:	2037	2041
Licensed Thermal Power (MWt):	2,775	2,775
Net Capacity (MWe):	851	860
Type of Reactor:	PWR	PWR
Nuclear Steam Supply System Vendor:	WEST	WEST

Cooling Water System

Type: Mechanical draft cooling towers
Source: Chattahoochee River
Source Temperature Range: 86°F (130°C) maximum
Condenser Flow Rate: 635,000 gpm (40.1 m^3/s) each unit
Design Condenser Temperature Rise: 20°F (11°C)
Intake Structure: Intake from river bank via storage pond
Discharge Structure: At river bank

Site Information

Total Area: 1,850 ac (749 ha)
Exclusion Distance: 0.78 mi (1.26 km)
Low Population Zone: 2 mi (3.22 km)
Nearest City: Columbus, Georgia; 2000 population: 185,781
Site Topography: Flat to rolling
Surrounding Area Topography: Rolling
Dominant Land Cover within 5 mi (8 km): Forest, agriculture, wetland
Level 1 Ecoregion within 5 mi (8 km): Eastern Temperate Forest
Level 3 Ecoregion within 5 mi (8 km): Southeastern Plains
Percent Wetland within 5 mi (8 km): 13.1, mostly freshwater forested/shrub wetland

Nearby Features: The nearest town is Columbia about 4 mi (6 km) N. Chattahoochee State Park is about 12 mi (19 km) S.
Population within a 50 mi (80 km) Radius: 393,639

FORT CALHOUN STATION

Location:	Washington County, Nebraska 19 mi (31 km) N of Omaha Latitude 41.5208°N; longitude 96.0767°W
Licensee:	Omaha Public Power District

Unit Information	Unit 1
Docket Number:	50-285
Construction Permit:	1968
Operating License:	1973
Commercial Operation:	1974
License Expiration:	2033
Licensed Thermal Power (MWt):	1,500
Net Capacity (MWe):	478
Type of Reactor:	PWR
Nuclear Steam Supply System Vendor:	CE

Cooling Water System

Type: Once-through
Source: Missouri River
Source Temperature Range: 0–27°C (32–80°F)
Condenser Flow Rate: 360,000 gpm (23 m^3/s)
Design Condenser Temperature Rise: 23°F (13°C)
Intake Structure: Concrete structure at river shore
Discharge Structure: At river shore

Site Information

Total Area: 660 ac (270 ha)
Exclusion Distance: 0.57 mi (0.92 km) minimum
Low Population Zone: 5 mi (8.05 km)
Nearest City: Omaha: 2000 population: 390,007
Site Topography: Flat to rolling
Surrounding Area Topography: Flat to rolling
Dominant Land Cover within 5 mi (8 km): Agriculture, herbaceous, wetland
Level 1 Ecoregion within 5 mi (8 km): Great Plains
Level 3 Ecoregion within 5 mi (8 km): Western Corn Belt Plains

Percent Wetland within 5 mi (8 km): 6.3, mostly lake; riverine; freshwater forested/shrub wetland
Nearby Features: The nearest town is De Soto 2 mi (3 km) SSE. De Soto National Wildlife Refuge is about 1 mi (1.6 km) E. Wilson Island State Park is about 4 mi (6 km) SE.
Population within a 50 mi (80 km) Radius: 852,717

GRAND GULF NUCLEAR STATION

Location: Clairborne County, Mississippi
25 mi (40 km) S of Vicksburg
Latitude 32.0075°N; longitude 91.0475°W
Licensee: Entergy Nuclear Operations, Inc.

Unit Information	Unit 1
Docket Number:	50-416
Construction Permit:	1974
Operating License:	1984
Commercial Operation:	1985
License Expiration:	2024
Licensed Thermal Power (MWt):	3,963
Net Capacity (MWe):	1,297
Type of Reactor:	BWR
Nuclear Steam Supply System Vendor:	GE

Cooling Water System

Type: Natural draft cooling towers
Source: Mississippi River
Source Temperature Range: 34–82°F (1–28°C)
Condenser Flow Rate: 572,000 gpm (36.1 m^3/s)
Design Condenser Temperature Rise: 30°F (17°C)
Intake Structure: A series of radial-collector wells along the shoreline
Discharge Structure: Discharge to river via a barge slip

Site Information

Total Area: 2,100 ac (850 ha)
Exclusion Distance: 0.43 mi (0.69 km) radius
Low Population Zone: 2 mi (3.22 km)
Nearest City: Jackson; 2000 population: 184,256
Site Topography: Flat to rolling
Surrounding Area Topography: Flat to rolling
Dominant Land Cover within 5 mi (8 km): Forest, wetland, open water
Level 1 Ecoregion within 5 mi (8 km): Eastern Temperate Forest
Level 3 Ecoregion within 5 mi (8 km): Mississippi Valley Loess Plains; Mississippi Alluvial Plain

Percent Wetland within 5 mi (8 km): 39.4, mostly freshwater forested/shrub wetland
Nearby Features: The nearest town is Grand Gulf 2 mi (3 km) N. The Natchez Trace Parkway is about 6 mi (10 km) SE. The Grand Gulf Military Park is just N of the site.
Population within a 50 mi (80 km) Radius: 357,525

H.B. ROBINSON STEAM ELECTRIC STATION

Location: Darlington County, South Carolina
26 mi (42 km) NE of Florence
Latitude 34.4025°N; longitude 80.1586°W
Licensee: Progress Energy

Unit Information	Unit 2
Docket Number:	50-261
Construction Permit:	1967
Operating License:	1970
Commercial Operation:	1971
License Expiration:	2030
Licensed Thermal Power (MWt):	2,339
Net Capacity (MWe):	710
Type of Reactor:	PWR
Nuclear Steam Supply System Vendor:	WEST

Cooling Water System

Type: Once-through, cooling pond
Source: Lake Robinson
Source Temperature Range: 46–85°F (8–29°C)
Condenser Flow Rate: 454,167 gpm (28.7 m^3/s)
Design Condenser Temperature Rise: 18°F (10°C)
Intake Structure: Concrete structure on edge of lake
Discharge Structure: 4.2 mi (6.8 km) canal discharging about 4 mi (6 km) upstream from intake

Site Information

Total Area: 6,020 ac (2,435 ha)
Exclusion Distance: 0.27 mi (0.43 km) radius
Low Population Zone: 4.5 mi (7.24 km)
Nearest City: Columbia; 2000 population: 116,278
Site Topography: Rolling
Surrounding Area Topography: Rolling
Dominant Land Cover within 5 mi (8 km): Forest, agriculture, herbaceous
Level 1 Ecoregion within 5 mi (8 km): Eastern Temperate Forest
Level 3 Ecoregion within 5 mi (8 km): Southeastern Plains
Percent Wetland within 5 mi (8 km): 13.5, mostly freshwater forested/shrub wetland

Nearby Features: The nearest town is Hartsville 5 mi (8 km) SE. Unit 1 is an adjacent 185 MWe capacity coal-fired plant. Sand Hills State Forest is about 4 mi (6 km) N. The Carolina Sandhills National Wildlife Refuge is about 5 mi (8 km) NNW.

Population within a 50 mi (80 km) Radius: 809,582

HOPE CREEK GENERATING STATION

Location:	Salem County, New Jersey 8 mi (13 km) SW of Salem Latitude 39.4678°N; longitude 75.5381°W
Licensee:	Public Service Electric and Gas Co.

Unit Information	Unit 1
Docket Number:	50-354
Construction Permit:	1974
Operating License:	1986
Commercial Operation:	1986
License Expiration:	2046
Licensed Thermal Power (MWt):	3,339
Net Capacity (MWe):	1,061
Type of Reactor:	BWR
Nuclear Steam Supply System Vendor:	GE

Cooling Water System

Type: Natural draft cooling tower
Source: Delaware River
Source Temperature Range: 34–81°F (1–27°C)
Condenser Flow Rate: 552,000 gpm (34.8 m^3/s)
Design Condenser Temperature Rise: 28°F (16°C)
Intake Structure: At edge of river
Discharge Structure: Pipe 10 ft (3 m) offshore

Site Information

Total Area: 740 ac (300 ha)
Exclusion Distance: 0.56 mi (0.90 km) radius
Low Population Zone: 5 mi (8.05 km) radius
Nearest City: Wilmington, Delaware; 2000 population: 72,664
Site Topography: Flat
Surrounding Area Topography: Flat
Dominant Land Cover within 5 mi (8 km): Open water, wetland, agriculture
Level 1 Ecoregion within 5 mi (8 km): Eastern Temperate Forest
Level 3 Ecoregion within 5 mi (8 km): Middle Atlantic Coastal Plain

Percent Wetland within 5 mi (8 km): 82.4, mostly estuarine and marine deepwater; estuarine and marine wetland

Nearby Features: The nearest town is Port Penn about 4 mi (6 km) NW in Delaware. The nearest railroad is 8 mi (13 km) NE. The plant is on the same site as the Salem Nuclear Generating Station.

Population within a 50 mi (80 km) Radius: 5,999,588

INDIAN POINT ENERGY CENTER

Location:	Westchester County, New York 24 mi (39 km) N of New York City Latitude 41.2714°N; longitude 73.9525°W
Licensee:	Entergy Nuclear Operations, Inc.

Unit Information	Unit 2	Unit 3
Docket Number:	50-247	50-286
Construction Permit:	1966	1969
Operating License:	1973	1976
Commercial Operation:	1974	1976
License Expiration:	2013	2015
Licensed Thermal Power (MWt):	3,216	3,216
Net Capacity (MWe):	1,020	1,025
Type of Reactor:	PWR	PWR
Nuclear Steam Supply System Vendor:	WEST	WEST

Cooling Water System

Type: Once-through
Source: Hudson River
Source Temperature Range: 32–78°F (0–26°C)
Condenser Flow Rate: 840,000 gal/min (53 m^3/s) each unit
Design Condenser Temperature Rise: 16.6°F (9.2°C)
Intake Structure: Concrete structure at river bank
Discharge Structure: Discharge canal to river exiting through 12 ports

Site Information

Total Area: 239 ac (96.7 ha)
Exclusion Distance: 0.20 mi (0.32 km) radius
Low Population Zone: 0.65 mi (1.05 km) radius
Nearest City: White Plains; 2000 population: 53,077
Site Topography: Hilly
Surrounding Area Topography: Hilly to mountainous
Dominant Land Cover within 5 mi (8 km): Forest, open water, developed: open space
Level 1 Ecoregion within 5 mi (8 km): Northern Forest
Level 3 Ecoregion within 5 mi (8 km): Northeastern Highlands
Percent Wetland within 5 mi (8 km): 19.0, mostly estuarine and marine deepwater

Nearby Features: The nearest town is Buchannan 2 mi (3 km) ESE. Camp Smith (military) is 1 mi (1.6 km) N, and West Point is 8 mi (13 km) N.
Population within a 50 mi (80 km) Radius: 16,791,654

KEWAUNEE POWER STATION

Location:	Kewaunee County, Wisconsin 27 mi (43 km) E of Green Bay Latitude 44.3431°N; longitude 87.5361°W
Licensee:	Dominion Generation

Unit Information	Unit 1
Docket Number:	50-305
Construction Permit:	1968
Operating License:	1973
Commercial Operation:	1974
License Expiration:	2033
Licensed Thermal Power (MWt):	1,772
Net Capacity (MWe):	556
Type of Reactor:	PWR
Nuclear Steam Supply System Vendor:	WEST

Cooling Water System

Type: Once-through
Source: Lake Michigan
Source Temperature Range: 34–67°F (1–19°C)
Condenser Flow Rate: 420,000 gpm (27 m^3/s)
Design Condenser Temperature Rise: 19°F (11°C)
Intake Structure: Intake crib 15 ft (4.6 km) deep 1,750 ft (533 m) from shore
Discharge Structure: At shoreline

Site Information

Total Area: 908 ac (367 ha)
Exclusion Distance: 0.75 mi (1.21 km)
Low Population Zone: 3 mi (4.83 km) radius
Nearest City: Green Bay; 2000 population: 102,313
Site Topography: Flat to rolling
Surrounding Area Topography: Flat to rolling
Dominant Land Cover within 5 mi (8 km): Open Water, Agriculture, Wetland
Level 1 Ecoregion within 5 mi (8 km): Eastern Temperate Forest
Level 3 Ecoregion within 5 mi (8 km): Southeastern Wisconsin Till Plains
Percent Wetland within 5 mi (8 km): 51.9, mostly lake

Nearby Features: The nearest town is Two Creeks about 3 mi (5 km) S. Point Beach Nuclear Plant is about 5 mi (8 km) S.

Population within a 50 mi (80 km) Radius: 1,585,415

LASALLE COUNTY STATION

Location: LaSalle County, Illinois
11 mi (18 km) SE of Ottawa
Latitude 41.2439°N; longitude 88.6708°W
Licensee: Exelon Generation Company

Unit Information	Unit 1	Unit 2
Docket Number:	50-373	50-374
Construction Permit:	1973	1973
Operating License:	1982	1984
Commercial Operation:	1984	1984
License Expiration:	2022	2023
Licensed Thermal Power (MWt):	3,489	3,489
Net Capacity (MWe):	1,118	1,120
Type of Reactor:	BWR	BWR
Nuclear Steam Supply System Vendor:	GE	GE

Cooling Water System

Type: Cooling pond
Source: Illinois River
Source Temperature Range: 47–85°F (8–29°C)
Condenser Flow Rate: 645,000 gpm (40.7 m^3/s) each unit
Design Condenser Temperature Rise: 24°F (13°C)
Intake Structure: Intake from 2,058 ac (832.8 ha) cooling pond, makeup from river
Discharge Structure: Discharge to cooling pond

Site Information

Total Area: 3,060 ac (1,240 ha)
Exclusion Distance: 0.32 mi (0.51 km)
Low Population Zone: 3.98 mi (6.41 km)
Nearest City: Joliet; 2000 population: 106,221
Site Topography: Flat
Surrounding Area Topography: Flat with hills along river
Dominant Land Cover within 5 mi (8 km): Agriculture, forest, open water
Level 1 Ecoregion within 5 mi (8 km): Eastern Temperate Forest
Level 3 Ecoregion within 5 mi (8 km): Central Corn Belt Plains
Percent Wetland within 5 mi (8 km): 4.9, mostly lake

Nearby Features: The nearest town is Seneca about 5 mi (8 km) NNE. Braidwood Station (nuclear plant) is about 20 mi (32 km) ENE, and Dresden Nuclear Power Station is about 22 mi (35 km) NE.

Population within a 50 mi (80 km) Radius: 1,498,644

LIMERICK GENERATING STATION

Location:	Montgomery County, Pennsylvania 21 mi (34 km) NW of Philadelphia Latitude 40.2200°N; longitude 75.5900°W
Licensee:	Exelon Generation Company

Unit Information	Unit 1	Unit 2
Docket Number:	50-352	50-353
Construction Permit:	1974	1974
Operating License:	1985	1989
Commercial Operation:	1986	1989
License Expiration:	2024	2029
Licensed Thermal Power (MWt):	3,458	3,458
Net Capacity (MWe):	1,134	1,134
Type of Reactor:	BWR	BWR
Nuclear Steam Supply System Vendor:	GE	GE

Cooling Water System

Type: Natural draft cooling towers
Source: Schuylkill River
Source Temperature Range: 42–82°F (6–28°C)
Condenser Flow Rate: 450,000 gpm (28 m^3/s) each unit
Design Condenser Temperature Rise: 30°F (17°C)
Intake Structure: Intake from river
Discharge Structure: Discharge to river

Site Information

Total Area: 595 ac (241 ha)
Exclusion Distance: 0.47 mi (0.76 km)
Low Population Zone: 1.30 mi (2.09 km)
Nearest City: Reading; 2000 population: 81,207
Site Topography: Rolling
Surrounding Area Topography: Rolling
Dominant Land Cover within 5 mi (8 km): Agriculture, forest, developed: high, medium, low density
Level 1 Ecoregion within 5 mi (8 km): Eastern Temperate Forest
Level 3 Ecoregion within 5 mi (8 km): Northern Piedmont

Percent Wetland within 5 mi (8 km): 2, mostly riverine
Nearby Features: The nearest town is Linfield about 1 mi (1.6 km) SE. Valley Forge State Park is 10 mi (16 km) SSE. U.S. Highway I-76 is about 10 mi (16 km) S.
Population within a 50 mi (80 km) Radius: 7,651,537

MCGUIRE NUCLEAR STATION

Location: Mecklenburg County, North Carolina
17 mi (27 km) NNW of Charlotte
Latitude 35.4322°N; longitude 80.9483°W

Licensee: Duke Energy Power Company

Unit Information	Unit 1	Unit 2
Docket Number:	50-369	50-370
Construction Permit:	1973	1973
Operating License:	1981	1983
Commercial Operation:	1981	1984
License Expiration:	2041	2043
Licensed Thermal Power (MWt):	3,411	3,411
Net Capacity (MWe):	1,100	1,100
Type of Reactor:	PWR	PWR
Nuclear Steam Supply System Vendor:	WEST	WEST

Cooling Water System

Type: Once-through
Source: Lake Norman
Source Temperature Range: 38–89°F (3–32°C)
Condenser Flow Rate: 1,756,944 gpm (111 m^3/s) both units
Design Condenser Temperature Rise: 22.1°F (12.3°C)
Intake Structure: Submerged and surface intakes at shoreline
Discharge Structure: 2,000 ft (610 m) discharge canal

Site Information

Total Area: 577 ac (234 ha)
Exclusion Distance: 0.47 mi (0.76 km) radius
Low Population Zone: 5.50 mi (8.85 km)
Nearest City: Charlotte; 2000 population: 540,828
Site Topography: Rolling
Surrounding Area Topography: Hilly
Dominant Land Cover within 5 mi (8 km): Forest, open water, agriculture
Level 1 Ecoregion within 5 mi (8 km): Eastern Temperate Forest
Level 3 Ecoregion within 5 mi (8 km): Piedmont
Percent Wetland within 5 mi (8 km): 21.4, mostly lake

Nearby Features: The nearest town is Lowesville about 3 mi (5 km) W. The dam forming Lake Norman and a hydroelectric power plant are adjacent to the site.
Population within a 50 mi (80 km) Radius: 2,425,097

MILLSTONE POWER STATION

Location: New London County, Connecticut
3 mi (5 km) WSW of New London
Latitude 41.3086°N; longitude 72.1681°W
Licensee: Dominion Generation

Unit Information	Unit 2	Unit 3
Docket Number:	50-336	50-423
Construction Permit:	1970	1974
Operating License:	1975	1986
Commercial Operation:	1975	1986
License Expiration:	2035	2045
Licensed Thermal Power (MWt):	2,700	3,650
Net Capacity (MWe):	884	1,227
Type of Reactor:	PWR	PWR
Nuclear Steam Supply System Vendor:	CE	WEST

Cooling Water System

Type: Once-through
Source: Long Island Sound
Source Temperature Range: 36–72°F (2–22°C)
Condenser Flow Rate: 1.46 million gpm (92 m^3/s) both units
Design Condenser Temperature Rise: 21°F (13°C) for Unit 2
17.5°F (9.7°C) for Unit 3
Intake Structure: On shore of Niantic Bay off Long Island Sound
Discharge Structure: Discharge to Niantic Bay via holding pond

Site Information

Total Area: 500 ac (200 ha)
Exclusion Distance: 0.34 mi (0.55 km) minimum
Low Population Zone: (2.40 mi 3.86 km) radius
Nearest City: New Haven; 2000 population: 123,626
Site Topography: Flat
Surrounding Area Topography: Flat to rolling
Dominant Land Cover within 5 mi (8 km): Open water, forest, developed: high, medium, low density
Level 1 Ecoregion within 5 mi (8 km): Eastern Temperate Forest

Level 3 Ecoregion within 5 mi (8 km): Northeastern Coastal Zone
Percent Wetland within 5 mi (8 km): 53.5, mostly estuarine and marine deepwater
Nearby Features: The nearest town is Niantic 2 mi (3 km) NW. U.S. Highway I-95 is about 4 mi (6 km) NNE. Stone Ranch Military Reservation is about 6 mi (10 km) NW. Harkness Memorial State Park, Bluff Point State Park, and Rocky Neck State Park are within 5 mi (8 km) of the site. The U.S. Department of Agriculture Plum Island facility is 10 mi (16 km) S in Long Island Sound. The decommissioned Haddam Neck Plant (nuclear) is 20 mi (32 km) NW.
Population within a 50 mi (80 km) Radius: 2,868,207

MONTICELLO NUCLEAR GENERATING PLANT

Location:	Wright County, Minnesota 35 mi (56 km) NW of Minneapolis Latitude 45.3333°N; longitude 93.8483°W
Licensee:	Northern States Power Company

Unit Information	Unit 1
Docket Number:	50-263
Construction Permit:	1967
Operating License:	1970
Commercial Operation:	1971
License Expiration:	2030
Licensed Thermal Power (MWt):	1,775
Net Capacity (MWe):	572
Type of Reactor:	BWR
Nuclear Steam Supply System Vendor:	GE

Cooling Water System

Type: Once-through and mechanical draft towers
Source: Mississippi River
Source Temperature Range: 32–85°F (0–29°C)
Condenser Flow Rate: 292,000 gpm (18 m^3/s)
Design Condenser Temperature Rise: 26.8°F (14.9°C)
Intake Structure: Canal
Discharge Structure: Canal

Site Information

Total Area: 2,150 ac (860 ha)
Exclusion Distance: 0.30 mi (0.48 km)
Low Population Zone: 1 mi (1.61 km)
Nearest City: Minneapolis; 2000 population: 382,618
Site Topography: Flat terraces
Surrounding Area Topography: Flat to gently sloping
Dominant Land Cover 5 mi within (8 km): Agriculture, forest, developed: open space
Level 1 Ecoregion within 5 mi (8 km): Eastern Temperate Forest
Level 3 Ecoregion within 5 mi (8 km): North Central Hardwood Forests

Percent Wetland within 5 mi (8 km): 11.8, mostly freshwater emergent wetland; lake; freshwater forested/shrub wetland

Nearby Features: The business district of Monticello is about 2 mi (3.2 km) SE. Sherburne National Wildlife Refuge is about 9 mi (14 km) N. Lake Maria State Park is about 6 mi (10 km) WSW, and Sand Dunes State Forest and campground are 9 mi (14 km) NE.

Population within a 50 mi (80 km) Radius: 2,740,995

NINE MILE POINT NUCLEAR STATION

Location:	Oswego County, New York 6 mi (10 km) NE of Oswego Latitude 43.5222°N; longitude 76.4100°W
Licensee:	Constellation Energy

Unit Information	Unit 1	Unit 2
Docket Number:	50-220	50-410
Construction Permit:	1965	1974
Operating License:	1968	1987
Commercial Operation:	1969	1988
License Expiration:	2029	2046
Licensed Thermal Power (MWt):	1,850	3,467
Net Capacity (MWe):	621	1,140
Type of Reactor:	BWR	BWR
Nuclear Steam Supply System Vendor:	GE	GE

Cooling Water System

Type: Unit 1: Once-through
Unit 2: Natural draft tower
Source: Lake Ontario
Source Temperature Range: 33–77°F (1–25°C)
Condenser Flow Rate: Unit 1: 290,278 gpm (18 m^3/s); Unit 2: 580,000 gpm (36.6 m^3/s)
Design Condenser Temperature Rise: Unit 1: 35°F (19.4°C);
Unit 2: 30°F (16.7°C)
Intake Structure: Unit 1: submerged pipeline about 850 ft (260 m) from shore;
Unit 2: submerged pipelines about 950 ft (300 m) and 1,050 ft (320 m) from shore
Discharge Structure: Diffuser pipe 555 ft (169 m) long serving both sides

Site Information

Total Area: 900 ac (360 ha)
Exclusion Distance: 1 mi (1.6 km) to the east, 0.87 mi (1.4 km) to the southwest, and 1.3 mi (2 km) to the southern site boundary
Low Population Zone: 4 mi (6.44 km) radius
Nearest City: Syracuse; 2000 population: 147,306
Site Topography: Flat to rolling

Surrounding Area Topography: Rolling
Dominant Land Cover within 5 mi (8 km): Open water, forest, agriculture
Level 1 Ecoregion within 5 mi (8 km): Eastern Temperate Forest
Level 3 Ecoregion within 5 mi (8 km): Eastern Great Lakes and Hudson Lowlands
Percent Wetland within 5 mi (8 km): 65.7, mostly lake
Nearby Features: The nearest town is Lakeview about 1 mi (1.6 km) WSW. Fort Ontario is about 6 mi (10 km) SW. James A. Fitzpatrick Nuclear Power Plant is 0.5 mi (0.8 km) E.
Population within a 50 mi (80 km) Radius: 914,668

NORTH ANNA POWER STATION

Location:	Louisa County, Virginia 40 mi (64 km) NW of Richmond Latitude 38.0608°N; longitude 77.7906°W
Licensee:	Dominion Generation

Unit Information	Unit 1	Unit 2
Docket Number:	50-338	50-339
Construction Permit:	1971	1971
Operating License:	1978	1980
Commercial Operation:	1978	1980
License Expiration:	2038	2040
Licensed Thermal Power (MWt):	2,893	2,893
Net Capacity (MWe):	981	973
Type of Reactor:	PWR	PWR
Nuclear Steam Supply System Vendor:	WEST	WEST

Cooling Water System

Type: Once-through
Source: Lake Anna
Source Temperature Range: 48–83°F (9–28°C)
Condenser Flow Rate: 1,900,000 gpm (120 m^3/s) both units
Design Condenser Temperature Rise: 14.5°F (8.1°C)
Intake Structure: Intake at lake shore
Discharge Structure: Discharged through lake via a 3,400 ac (1,400 ha) cooling pond

Site Information

Total Area: 18,643 ac (7,550 ha)
Exclusion Distance: 0.84 mi (1.35 km)
Low Population Zone: 9.66 km (6 mi)
Nearest City: Richmond; 2000 population: 197,790
Site Topography: Rolling
Surrounding Area Topography: Rolling
Dominant Land Cover within 5 mi (8 km): Agriculture, forest, agriculture, open water
Level 1 Ecoregion within 5 mi (8 km): Eastern Temperate Forest
Level 3 Ecoregion within 5 mi (8 km): Piedmont
Percent Wetland within 5 mi (8 km): 21.1, mostly lake

Nearby Features: The nearest town is Centreville 1 mi (1.6 km) SW. Fredericksburg and Spotsylvania National Military Park is about 15 mi (24 km) NE.
Population within a 50 mi (80 km) Radius: 1,614,983

OCONEE NUCLEAR STATION

Location: Oconee County, South Carolina
26 mi (42 km) W of Greenville
Latitude 34.7917°N; longitude 82.8986°W
Licensee: Duke Energy Power Company

Unit Information	Unit 1	Unit 2	Unit 3
Docket Number:	50-269	50-270	50-287
Construction Permit:	1967	1967	1967
Operating License:	1973	1973	1974
Commercial Operation:	1973	1974	1974
License Expiration:	2033	2033	2034
Licensed Thermal Power (MWt):	2,568	2,568	2,568
Net Capacity (MWe):	846	846	846
Type of Reactor:	PWR	PWR	PWR
Nuclear Steam Supply System Vendor:	B&W	B&W	B&W

Cooling Water System

Type: Once-through
Source: Lake Keowee
Source Temperature Range: 44–77°F (7–25°C)
Condenser Flow Rate: 1,527,778 gpm (96 m^3/s) all units
Design Condenser Temperature Rise: 17.2°F (9.6°C)
Intake Structure: A skimmer wall draws water from the depths of 735 ft (223 m).
Discharge Structure: All three units discharge through one structure near the Keowee Dam.

Site Information

Total Area: 510 ac (210 ha)
Exclusion Distance: 1 mi (1.6 km) radius
Low Population Zone: 6 mi (9.66 km)
Nearest City: Greenville; 2000 population: 56,002
Site Topography: Flat to rolling
Surrounding Area Topography: Hilly
Dominant Land Cover within 5 mi (8 km): Forest, open water, agriculture
Level 1 Ecoregion within 5 mi (8 km): Eastern Temperate Forest
Level 3 Ecoregion within 5 mi (8 km): Piedmont
Percent Wetland within 5 mi (8 km): 22.3, mostly lake

Nearby Features: The nearest town is Six Mile (6 4 mi km) ENE. Keowee Dam is close to the plant. Chattahoochee National Forest is about 15 mi (24 km) W.

Population within a 50 mi (80 km) Radius: 1,226,479

OYSTER CREEK NUCLEAR GENERATING STATION

Location: Ocean County, New Jersey
9 mi (14 km) S of Toms River
Latitude 39.8142°N; longitude 74.2064°W
Licensee: Exelon Generation Company

Unit Information	Unit 1
Docket Number:	50-219
Construction Permit:	1964
Operating License:	1969
Commercial Operation:	1969
License Expiration:	2029
Licensed Thermal Power (MWt):	1,930
Net Capacity (MWe):	619
Type of Reactor:	BWR
Nuclear Steam Supply System Vendor:	GE

Cooling Water System

Type: Once-through
Source: Barnegat Bay
Source Temperature Range: 35–75°F (2–24°C)
Condenser Flow Rate: 460,000 gpm (29 m^3/s)
Design Condenser Temperature Rise: 14°F (8°C)
Intake Structure: Forked River serves as a canal for intake and discharge to Barnegat Bay.
Discharge Structure: Forked River serves as a canal for intake and discharge to Barnegat Bay.

Site Information

Total Area: 800 ac (323.8 ha)
Exclusion Distance: 0.25 mi (0.40 km)
Low Population Zone: 2 mi (3.22 km)
Nearest City: Atlantic City; 2000 population: 40,517
Site Topography: Flat
Surrounding Area Topography: Rolling plains to flat lowlands
Dominant Land Cover within 5 mi (8 km): Forest, open water, developed: high, medium, low density
Level 1 Ecoregion within 5 mi (8 km): Eastern Temperate Forest
Level 3 Ecoregion within 5 mi (8 km): Atlantic Coastal Pine Barrens

Percent Wetland within 5 mi (8 km): 45, mostly estuarine and marine deepwater; freshwater forested/shrub wetland

Nearby Features: The nearest town is Forked River about 2 mi (3 km) N. The Garden State Parkway is 1 mi (1.6 km) W. There is a large influx of recreationists and tourists in the summer.

Population within a 50 mi (80 km) Radius: 4,243,462

PALISADES NUCLEAR PLANT

Location:	Van Buren County, Michigan 35 mi (56 km) W of Kalamazoo Latitude 42.3222°N; longitude 86.3153°W
Licensee:	Entergy Nuclear Operations, Inc.

Unit Information	Unit 1
Docket Number:	50-255
Construction Permit:	1967
Operating License:	1972
Commercial Operation:	1973
License Expiration:	2031
Licensed Thermal Power (MWt):	2,565
Net Capacity (MWe):	778
Type of Reactor:	PWR
Nuclear Steam Supply System Vendor:	CE

Cooling Water System

Type: Mechanical draft cooling towers
Source: Lake Michigan
Source Temperature Range: 35–75°F (2–24°C)
Condenser Flow Rate: 98,000 gpm (6.2 m^3/s)
Design Condenser Temperature Rise: 25°F (14°C)
Intake Structure: Intake crib 3,300 ft (1,000 m) from shore
Discharge Structure: 108 ft (33 m) long canal

Site Information

Total Area: 432 ac (174.8 ha)
Exclusion Distance: 0.44 mi (0.71 km) radius
Low Population Zone: Not available
Nearest City: Kalamazoo; 2000 population: 77,145
Site Topography: Flat to rolling
Surrounding Area Topography: Rolling
Dominant Land Cover within 5 mi (8 km): Open water, forest, agriculture
Level 1 Ecoregion within 5 mi (8 km): Eastern Temperate Forest
Level 3 Ecoregion within 5 mi (8 km): S. Michigan/N. Indiana Drift Plains
Percent Wetland within 5 mi (8 km): 58.1, mostly lake

Nearby Features: The nearest town is South Haven about 4 mi (6 km) N. Van Buren State Park joins the plant on the north. Many tourists come to the beaches in the summer. The C&O Railway is about 2 mi (3 km) E. Highway I-196 is about 1 mi (1.6 km) E.

Population within a 50 mi (80 km) Radius: 1,287,558

PALO VERDE NUCLEAR GENERATING STATION

Location: Maricopa County, Arizona
34 mi (55 km) W of Phoenix
Latitude 33.3881°N; longitude 112.8644°W

Licensee: Arizona Public Service Co.

Unit Information	Unit 1	Unit 2	Unit 3
Docket Number:	50-528	50-529	50-530
Construction Permit:	1976	1976	1976
Operating License:	1985	1986	1987
Commercial Operation:	1986	1986	1988
License Expiration:	2045	2046	2047
Licensed Thermal Power (MWt):	3,990	3,990	3,990
Net Capacity (MWe):	1,335	1,335	1,335
Type of Reactor:	PWR	PWR	PWR
Nuclear Steam Supply System Vendor:	CE	CE	CE

Cooling Water System

Type: Mechanical draft cooling towers treatment plant
Source: Phoenix City Sewage
Source Temperature Range: Not available
Condenser Flow Rate: 560,000 gpm (35 m^3/s) each unit
Design Condenser Temperature Rise: 32.1°F (17.8°C)
Intake Structure: 35 mi (56 km) underground pipeline from Phoenix 91st Avenue Sewage Treatment Plant
Discharge Structure: Blowdown from the circulating water system is directed to onsite evaporation ponds without requiring any offsite discharge

Site Information

Total Area: 4,050 ac (1,640 ha)
Exclusion Distance: 0.54 mi (0.87 km) minimum
Low Population Zone: 4 mi (6.44 km) radius
Nearest City: Phoenix; 2000 population: 1,321,045
Site Topography: Flat with hills
Surrounding Area Topography: Flat with hills
Dominant Land Cover within 5 mi (8 km): Shrub/scrub, agriculture, developed: open space
Level 1 Ecoregion within 5 mi (8 km): North American Desert

Level 3 Ecoregion within 5 mi (8 km): Sonoran Basin and Range
Percent Wetland within 5 mi (8 km): 1.2, mostly lake
Nearby Features: The nearest town is Wintersburg about 3 mi (5 km) N. U.S. Highway I-10 is about 7 mi (11 km) N. The Southern Pacific Railroad is about 5 mi (8 km) SE.
Population within a 50 mi (80 km) Radius: 1,781,095

PEACH BOTTOM ATOMIC POWER STATION

Location:	York County, Pennsylvania 18 mi (29 km) S of Lancaster Latitude 39.7589°N; longitude 76.2692°W
Licensee:	Exelon Generation Company

Unit Information	Unit 2	Unit 3
Docket Number:	50-277	50-278
Construction Permit:	1968	1968
Operating License:	1973	1974
Commercial Operation:	1974	1974
License Expiration:	2033	2034
Licensed Thermal Power (MWt):	3,514	3,514
Net Capacity (MWe):	1,112	1,112
Type of Reactor:	BWR	BWR
Nuclear Steam Supply System Vendor:	GE	GE

Cooling Water System

Type: Once-through, with helper mechanical draft towers
Source: Conowingo Pond
Source Temperature Range: 34–80°F (1–27°C)
Condenser Flow Rate: 1.5 million gpm (95 m^3/s) (both units)
Design Condenser Temperature Rise: 20.8°F (11.5°C)
Intake Structure: Intake from Conowingo Pond through a small intake pond
Discharge Structure: 5,000 ft (1,520 m) canal to Conowingo Pond

Site Information

Total Area: 620 ac (248 ha)
Exclusion Distance: 0.51 mi (0.82 km)
Low Population Zone: 1.38 mi (2.22 km)
Nearest City: Lancaster; 2000 population: 56,348
Site Topography: Rolling to hilly
Surrounding Area Topography: Rolling to hilly
Dominant Land Cover within 5 mi (8 km): Agriculture, forest, open water
Level 1 Ecoregion within 5 mi (8 km): Eastern Temperate Forest
Level 3 Ecoregion within 5 mi (8 km): Northern Piedmont
Percent Wetland within 5 mi (8 km): 14.5, mostly lake

Nearby Features: The nearest town is Slate Hill 2 mi (3 km) SW. Susquehanna State Park is about 3 mi (5 km) N. U.S. Highway I-95 is about 15 mi (24 km) SE. Conowingo Dam, about 8 mi (13 km) SE on the Susquehanna River, forms Conowingo Pond. Unit 1 is a 40 MWe nuclear plant on the same site and was retired from service in 1974. Three Mile Island Nuclear Station is 35 mi (56 km) upstream on the Susquehanna River.

Population within a 50 mi (80 km) Radius: 5,270,600

PERRY NUCLEAR POWER PLANT

Location: Lake County, Ohio
7 mi (11 km) NE of Painesville
Latitude 41.8008°N; longitude 81.1442°W
Licensee: FirstEnergy Nuclear Operating Co.

Unit Information	Unit 1
Docket Number:	50-440
Construction Permit:	1977
Operating License:	1986
Commercial Operation:	1987
License Expiration:	2026
Licensed Thermal Power (MWt):	3,758
Net Capacity (MWe):	1,261
Type of Reactor:	BWR
Nuclear Steam Supply System Vendor:	GE

Cooling Water System

Type: Natural draft cooling tower
Source: Lake Erie
Source Temperature Range: 32–79°F (0–26°C)
Condenser Flow Rate: 545,400 gpm (34.41 m^3/s)
Design Condenser Temperature Rise: 32°F (18°C)
Intake Structure: Submerged multiport structure 2,550 ft (777 m) offshore
Discharge Structure: Submerged diffuser 1,650 ft (503 m) offshore

Site Information

Total Area: 1,100 ac (450 ha)
Exclusion Distance: 0.55 mi (0.89 km) radius
Low Population Zone: 2.50 mi (4.02 km)
Nearest City: Euclid; 2000 population: 52,717
Site Topography: Flat
Surrounding Area Topography: Rolling
Dominant Land Cover within 5 mi (8 km): Open water, forest, developed: high, medium, low density
Level 1 Ecoregion within 5 mi (8 km): Eastern Temperate Forest

Level 3 Ecoregion within 5 mi (8 km): Eastern Great Lakes and Hudson Lowlands; Erie Drift Plain
Percent Wetland within 5 mi (8 km): 49.3, mostly lake
Nearby Features: The nearest town is North Perry 1 mi (1.6 km) SW. The Penn Central Railroad is about 3 mi (5 km) S. U.S. Highway I-90 is about 5 mi (8 km) S.
Population within a 50 mi (80 km) Radius: 4,923,662

PILGRIM NUCLEAR POWER STATION

Location:	Plymouth County, Massachusetts 4 mi (6 km) SE of Plymouth Latitude 41.9444°N; longitude 70.5794°W
Licensee:	Entergy Nuclear Operations, Inc.

Unit Information	Unit 1
Docket Number:	50-293
Construction Permit:	1968
Operating License:	1972
Commercial Operation:	1972
License Expiration:	2012
Licensed Thermal Power (MWt):	2,028
Net Capacity (MWe):	685
Type of Reactor:	BWR
Nuclear Steam Supply System Vendor:	GE

Cooling Water System

Type: Once-through
Source: Cape Cod Bay
Source Temperature Range: 35.6–71.6°F (2–22°C)
Condenser Flow Rate: 311,100 gpm (19.6 m^3/s)
Design Condenser Temperature Rise: 32°F (18°C)
Intake Structure: Concrete structure at edge of bay protected by a breakwater
Discharge Structure: 850 ft (260 m) long canal

Site Information

Total Area: 140 ac (57 ha)
Exclusion Distance: 0.33 mi (0.53 km)
Low Population Zone: 4.20 mi (6.76 km)
Nearest City: Brockton; 2000 population: 94,304
Site Topography: Flat to rolling
Surrounding Area Topography: Rolling to hilly
Dominant Land Cover within 5 mi (8 km): Open water, forest, developed: high, medium, low density
Level 1 Ecoregion within 5 mi (8 km): Eastern Temperate Forest
Level 3 Ecoregion within 5 mi (8 km): Atlantic Coastal Pine Barrens; Northeastern Coastal Zone

Percent Wetland within 5 mi (8 km): 64.4, mostly estuarine and marine deepwater
Nearby Features: The nearest town is Plymouth about 4 mi (6 km) NW. Miles Standish State Forest is about 6 mi (10 km) SW. Plymouth Rock and Plymouth Plantation historical sites are about 5 mi (8 km) W.
Population within a 50 mi (80 km) Radius: 4,629,116

POINT BEACH NUCLEAR PLANT

Location:	Manitowoc County, Wisconsin 13 mi (21 km) NNW of Manitowoc Latitude 44.2808°N; longitude 87.5361°W
Licensee:	Florida Power & Light Co.

Unit Information	Unit 1	Unit 2
Docket Number:	50-266	50-301
Construction Permit:	1967	1968
Operating License:	1970	1972
Commercial Operation:	1970	1972
License Expiration:	2030	2033
Licensed Thermal Power (MWt):	1,540	1,540
Net Capacity (MWe):	512	514
Type of Reactor:	PWR	PWR
Nuclear Steam Supply System Vendor:	WEST	WEST

Cooling Water System

Type: Once-through
Source: Lake Michigan
Source Temperature Range: Not available
Condenser Flow Rate: 350,000 gpm (22 m^3/s) each unit
Design Condenser Temperature Rise: 19.3°F (10.7°C)
Intake Structure: Submerged structure 1,750 ft (533 m) from shore
Discharge Structure: 2 steel piling troughs, extending 200 ft (61 m) into Lake Michigan

Site Information

Total Area: 1,260 ac (510 ha)
Exclusion Distance: 0.74 mi (1.19 km) radius
Low Population Zone: 5.60 mi (9.01 km)
Nearest City: Green Bay; 2000 population: 102,313
Site Topography: Flat to rolling
Surrounding Area Topography: Rolling
Dominant Land Cover within 5 mi (8 km): Open water, agriculture, wetland
Level 1 Ecoregion within 5 mi (8 km): Eastern Temperate Forest
Level 3 Ecoregion within 5 mi (8 km): Southeastern Wisconsin Till Plains
Percent Wetland within 5 mi (8 km): 54, mostly lake

Nearby Features: The nearest town is Two Creeks 1 mi (1.6 km) NNW. Point Beach State Forest is just S of the site. The Kewaunee Nuclear Power Plant is about 5 mi (8 km) N.

Population within a 50 mi (80 km) Radius: 1,622,052

PRAIRIE ISLAND NUCLEAR GENERATING PLANT

Location:	Goodhue County, Minnesota 28 mi (45 km) SE of Minneapolis Latitude 44.6219°N; longitude 92.6331°W
Licensee:	Northern States Power Co.

Unit Information	Unit 1	Unit 2
Docket Number:	50-282	50-306
Construction Permit:	1968	1968
Operating License:	1973	1974
Commercial Operation:	1973	1974
License Expiration:	2033	2034
Licensed Thermal Power (MWt):	1,650	1,650
Net Capacity (MWe):	551	545
Type of Reactor:	PWR	PWR
Nuclear Steam Supply System Vendor:	WEST	WEST

Cooling Water System

Type: Once-through and/or mechanical draft cooling towers
Source: Mississippi River
Source Temperature Range: 32–82°F (0–28°C)
Condenser Flow Rate: 294,000 gpm (18.6 m^3/s) each unit
Design Condenser Temperature Rise: 27°F (15°C)
Intake Structure: Short canal
Discharge Structure: Discharges to a basin then to towers and/or river

Site Information

Total Area: 560 ac (230 ha)
Exclusion Distance: 0.43 mi (0.69 km) radius
Low Population Zone: 1.50 mi (2.41 km)
Nearest City: Minneapolis; 2000 population: 382,618
Site Topography: Flat to rolling
Surrounding Area Topography: Rolling
Dominant Land Cover within 5 mi (8 km): Agriculture, forest, wetland
Level 1 Ecoregion within 5 mi (8 km): Eastern Temperate Forest
Level 3 Ecoregion within 5 mi (8 km): Driftless Area
Percent Wetland within 5 mi (8 km): 31.9, mostly freshwater forested/shrub wetland; lake

Nearby Features: The business district of the town of Red Wing is 6 mi (9.6 km) SE. A railroad line is just SW of the site.
Population within a 50 mi (80 km) Radius: 2,731,953

QUAD CITIES NUCLEAR POWER STATION

Location:	Rock Island County, Illinois 20 mi (32 km) NE of Moline Latitude 41.7261°N; longitude 90.3100°W
Licensee:	Exelon Generation Co.

Unit Information	Unit 1	Unit 2
Docket Number:	50-254	50-265
Construction Permit:	1967	1967
Operating License:	1972	1972
Commercial Operation:	1973	1973
License Expiration:	2032	2032
Licensed Thermal Power (MWt):	2,957	2,957
Net Capacity (MWe):	867	869
Type of Reactor:	BWR	BWR
Nuclear Steam Supply System Vendor:	GE	GE

Cooling Water System

Type: Once-through
Source: Mississippi River
Source Temperature Range: 32–85°F (0–29°C)
Condenser Flow Rate: 970,000 gpm (61 m^3/s) both units
Design Condenser Temperature Rise: 28°F (15.6°C)
Intake Structure: Canal at edge of river
Discharge Structure: Two-pipe diffuser system on bottom of river

Site Information

Total Area: (817 ac 331 ha)
Exclusion Distance: 0.50 mi (0.80 km)
Low Population Zone: 3 mi (4.83 km)
Nearest City: Davenport, Iowa; 2000 population: 98,359
Site Topography: Flat
Surrounding Area Topography: Flat
Dominant Land Cover within 5 mi (8 km): Agriculture, wetland, forest
Level 1 Ecoregion within 5 mi (8 km): Eastern Temperate Forest
Level 3 Ecoregion within 5 mi (8 km): Interior River Valley and Hills; Western Corn Belt Plains
Percent Wetland within 5 mi (8 km): 22.2, mostly freshwater forested/shrub wetland; lake

Nearby Features: The nearest town is Folletts 3 mi (5 km) NW. The Rock Island Railroad is 2 mi (3 km) W and the Chicago, Milwaukee, and St. Paul Railroad is 1 mi (1.6 km) E. The Rock Island Arsenal is about 15 mi (24 km) SW.
Population within a 50 mi (80 km) Radius: 656,527

R.E. GINNA NUCLEAR POWER PLANT

Location:	Wayne County, New York 20 mi (32 km) NE of Rochester Latitude 43.2778°N; longitude 77.3089°W
Licensee:	Constellation Energy

Unit Information	Unit 1
Docket Number:	50-244
Construction Permit:	1966
Operating License:	1969
Commercial Operation:	1970
License Expiration:	2029
Licensed Thermal Power (MWt):	1,775
Net Capacity (MWe):	498
Type of Reactor:	PWR
Nuclear Steam Supply System Vendor:	WEST

Cooling Water System

Type: Once-through
Source: Lake Ontario
Source Temperature Range: 32–80°F (0–27°C)
Condenser Flow Rate: 340,000 gpm (21.4 m^3/s)
Design Condenser Temperature Rise: 20°F (11°C)
Intake Structure: 3,100 ft (945 m) from shore, at a depth of 33 ft (10 m)
Discharge Structure: Canal discharges to Lake Ontario at shoreline.

Site Information

Total Area: 488 ac (197 ha)
Exclusion Distance: 0.29–0.85 mi (0.47–1.38 km)
Low Population Zone: 3 mi (4.83 km)
Nearest City: Rochester; 2000 population: 219,773
Site Topography: Gently rolling to flat
Surrounding Area Topography: Sloping
Dominant Land Cover within 5 mi (8 km): Open water, agriculture, forest
Level 1 Ecoregion within 5 mi (8 km): Eastern Temperate Forest
Level 3 Ecoregion within 5 mi (8 km): Eastern Great Lakes and Hudson Lowlands
Percent Wetland within 5 mi (8 km): 63.1, mostly lake

Nearby Features: The nearest town is Lakeside 2 mi (3 km) SW. The N.Y. Central Railroad is about 3 m (5 km) S.

Population within a 50 mi (80 km) Radius: 1,250,000

RIVER BEND STATION

Location:	West Feliciana County, Louisiana 24 mi (39 km) NNW of Baton Rouge Latitude 30.7569°N; longitude 91.3314°W
Licensee:	Entergy Nuclear Operations, Inc.

Unit Information	Unit 1
Docket Number:	50-458
Construction Permit:	1977
Operating License:	1985
Commercial Operation:	1986
License Expiration:	2025
Licensed Thermal Power (MWt):	3,091
Net Capacity (MWe):	989
Type of Reactor:	BWR
Nuclear Steam Supply System Vendor:	GE

Cooling Water System

Type: Mechanical draft cooling towers
Source: Mississippi River
Source Temperature Range: Not available
Condenser Flow Rate: 508,470 gpm (32.08 m^3/s)
Design Condenser Temperature Rise: 27°F (15°C)
Intake Structure: At river bank
Discharge Structure: Pipe extending into the river

Site Information

Total Area: 3,342 ac (1,352 ha)
Exclusion Distance: 0.57 mi (0.92 km) radius
Low Population Zone: 2.50 mi (4.02 km) radius
Nearest City: Baton Rouge; 2000 population: 227,818
Site Topography: Flat
Surrounding Area Topography: Flat to rolling
Dominant Land Cover within 5 mi (8 km): Wetland, forest, agriculture
Level 1 Ecoregion within 5 mi (8 km): Eastern Temperate Forest
Level 3 Ecoregion within 5 mi (8 km): Mississippi Valley Loess Plains; Mississippi Alluvial Plain
Percent Wetland within 5 mi (8 km): 41.6, mostly freshwater forested/shrub wetland

Nearby Features: The nearest town is St. Francisville 3 mi (5 km) NW. Audubon Memorial State Park is about 3 mi (5 km) NNE. The Illinois Central Railroad crosses the site.

Population within a 50 mi (80 km) Radius: 866,314

SAINT LUCIE NUCLEAR PLANT

Location: St. Lucie County, Florida
7 mi (11 km) SE of Fort Pierce
Latitude 27.3486°N; longitude 80.2464°W
Licensee: Florida Power & Light Co.

Unit Information	Unit 1	Unit 2
Docket Number:	50-335	50-389
Construction Permit:	1970	1977
Operating License:	1976	1983
Commercial Operation:	1976	1983
License Expiration:	2036	2043
Licensed Thermal Power (MWt):	2,700	2,700
Net Capacity (MWe):	839	839
Type of Reactor:	PWR	PWR
Nuclear Steam Supply System Vendor:	CE	CE

Cooling Water System

Type: Once-through
Source: Atlantic Ocean
Source Temperature Range: 87°F (31°C)
Condenser Flow Rate: 968,000 gpm (61 m^3/s) both units
Design Condenser Temperature Rise: 24°F (13°C).
Intake Structure: 1,200 ft (370 m) offshore
Discharge Structure: Unit 1 is 1,500 ft (460 m) offshore; Unit 2 is a multisport discharge 3,400 ft (1,040 m) offshore

Site Information

Total Area: 1,130 ac (457 ha)
Exclusion Distance: 0.97 mi (1.56 km) radius
Low Population Zone: 1 mi (1.61 km)
Nearest City: West Palm Beach; 2000 population: 82,103
Site Topography: Flat land and water
Surrounding Area Topography: Flat
Dominant Land Cover within 5 mi (8 km): Open water, wetland, developed: high, medium, low density
Level 1 Ecoregion within 5 mi (8 km): Eastern Temperate Forest

Level 3 Ecoregion within 5 mi (8 km): Southern Coastal Plain
Percent Wetland within 5 mi (8 km): 77.8, mostly estuarine and marine deepwater
Nearby Features: The nearest town is Ankona 2 mi (3 km) W. The Florida East Coast Railroad is about 2 mi (3 km) W. The plant is on Hutchinson Island, which is separated from the mainland by the Indian River, which is part of the Intracoastal Waterway. A causeway to the mainland is about 6 mi (10 km) SSE.
Population within a 50 mi (80 km) Radius: 1,180,000

SALEM NUCLEAR GENERATING STATION

Location:	Salem County, New Jersey 8 mi (13 km) SW of Salem Latitude 39.4628°N; longitude 75.5358°W
Licensee:	Public Service Electric and Gas Co.

Unit Information	Unit 1	Unit 2
Docket Number:	50-272	50-311
Construction Permit:	1968	1968
Operating License:	1976	1981
Commercial Operation:	1977	1981
License Expiration:	2036	2040
Licensed Thermal Power (MWt):	3,459	3,459
Net Capacity (MWe):	1,174	1,130
Type of Reactor:	PWR	PWR
Nuclear Steam Supply System Vendor:	WEST	WEST

Cooling Water System

Type: Once-through
Source: Delaware River
Source Temperature Range: 33–79°F (1–26°C)
Condenser Flow Rate: 1,100,000 gpm (69 m^3/s) each unit
Design Condenser Temperature Rise: 13.6°F (7.6°C)
Intake Structure: 12-bay structure on edge of river
Discharge Structure: Submerged pipes extending 500 ft (150 m) into the river

Site Information

Total Area: 700 ac (280 ha)
Exclusion Distance: 0.80 mi (1.29 km)
Low Population Zone: 5 mi (8.05 km)
Nearest City: Wilmington, Delaware; 2000 population: 72,664
Site Topography: Flat
Surrounding Area Topography: Flat
Dominant Land Cover within 5 mi (8 km): Open water, wetland, agriculture
Level 1 Ecoregion within 5 mi (8 km): Eastern Temperate Forest
Level 3 Ecoregion within 5 mi (8 km): Middle Atlantic Coastal Plain

Percent Wetland within 5 mi (8 km): 84, mostly estuarine and marine deepwater; estuarine and marine wetland

Nearby Features: The nearest town is Port Penn about 4 mi (6 km) NW in Delaware. The nearest railroad is 8 mi (13 km) NE. The plant is on the same site as the Hope Creek Generating Station (nuclear).

Population within a 50 mi (80 km) Radius: 5,975,864

SAN ONOFRE NUCLEAR GENERATING STATION

Location: San Diego County, California
5 mi (8 km) SE of San Clemente
Latitude 33.3703°N; longitude 117.5569°W
Licensee: Southern California Edison Co.

Unit Information	Unit 2	Unit 3
Docket Number:	50-361	50-362
Construction Permit:	1973	1973
Operating License:	1982	1983
Commercial Operation:	1983	1984
License Expiration:	2022	2022
Licensed Thermal Power (MWt):	3,438	3,438
Net Capacity (MWe):	1,070	1,080
Type of Reactor:	PWR	PWR
Nuclear Steam Supply System Vendor:	CE	CE

Cooling Water System

Type: Once-through
Source: Pacific Ocean
Source Temperature Range: 54–73°F (12–23°C)
Condenser Flow Rate: Unit 2: 797,000 gpm (50.3 m^3/s)
Design Condenser Temperature Rise: 20°F (11°C)
Intake Structure: Velocity-cap structure about 3,400 ft (1,040 m) from shore in water 30 ft (9 m) deep
Discharge Structure: Diffuser port systems extending 3,800–8,500 ft (1,160–2,590 m) from shore

Site Information

Total Area: 84 ac (34 ha)
Exclusion Distance: 0.37 mi (0.60 km)
Low Population Zone: 1.95 mi (3.14 km)
Nearest City: Oceanside; 2000 population: 161,029
Site Topography: Narrow, sloping coastal plain and sea cliffs
Surrounding Area Topography: Hilly
Dominant Land Cover within 5 mi (8 km): Open water, shrub/scrub, developed: high, medium, low density

Level 1 Ecoregion within 5 mi (8 km): Mediterranean California
Level 3 Ecoregion within 5 mi (8 km): Southern and Central California Chaparral and Oak Woodlands
Percent Wetland within 5 mi (8 km): 50.7, mostly estuarine and marine deepwater
Nearby Features: The nearest town is San Clemente 5 mi (8 km) NW. The site is surrounded by Camp Pendleton Marine Base. Camps on the base are 1.5 mi (2.4 km) or more from the site. U.S. Highway I-5 and the Atchison, Topeka, and Santa Fe Railroad are adjacent to the site to the east.
Population within a 50 mi (80 km) Radius: 12,404,757

SEABROOK STATION

Location:	Rockingham County, New Hampshire 13 mi (21 km) SSW of Portsmouth Latitude 42.8983°N; longitude 70.8497°W
Licensee:	Florida Power & Light Company

Unit Information	Unit 1
Docket Number:	50-443
Construction Permit:	1976
Operating License:	1990
Commercial Operation:	1990
License Expiration:	2030
Licensed Thermal Power (MWt):	3,648
Net Capacity (MWe):	1,295
Type of Reactor:	PWR
Nuclear Steam Supply System Vendor:	WEST

Cooling Water System

Type: Once-through
Source: Atlantic Ocean
Source Temperature Range: 37–55°F (3–13°C)
Condenser Flow Rate: 399,000 gpm (25.2 m^3/s)
Design Condenser Temperature Rise: 38°F (21°C)
Intake Structure: 3 structures 50 ft (15 m) below sea level with pipeline submerged about 175 ft (50 m) below mean sea level and extending about 7,000 ft (2,100 m) offshore
Discharge Structure: Submerged pipeline ending in a diffuser located about 5,500 ft (1,675 m) offshore and about 5,000 ft (1,525 m) S of intake

Site Information

Total Area: 896 ac (363 ha)
Exclusion Distance: 0.57 mi (0.92 km) minimum
Low Population Zone: 1.25 mi (2.01 km)
Nearest City: Lawrence, Massachusetts; 2000 population: 72,043
Site Topography: Flat
Surrounding Area Topography: Flat to rolling
Dominant Land Cover within 5 mi (8 km): Open water, forest, developed: high, medium, low density

Level 1 Ecoregion within 5 mi (8 km): Eastern Temperate Forest
Level 3 Ecoregion within 5 mi (8 km): Northeastern Coastal Zone
Percent Wetland within 5 mi (8 km): 45.2, mostly estuarine and marine deepwater; estuarine and marine wetland
Nearby Features: The nearest town is Seabrook 1 mi (1.6 km) W. U.S. Highway I-95 is about 1 mi (1.6 km) W. The Boston and Maine Railroad is adjacent to the site. Hampton Beach State Park is 2 mi (3 km) E.
Population within a 50 mi (80 km) Radius: 6,932,660

SEQUOYAH NUCLEAR PLANT

Location: Hamilton County, Tennessee
10 mi (16 km) NE of Chattanooga
Latitude 35.2233°N; longitude 85.0878°W
Licensee: Tennessee Valley Authority

Unit Information	Unit 1	Unit 2
Docket Number:	50-327	50-328
Construction Permit:	1970	1970
Operating License:	1980	1981
Commercial Operation:	1981	1982
License Expiration:	2020	2021
Licensed Thermal Power (MWt):	3,455	3,455
Net Capacity (MWe):	1,148	1,126
Type of Reactor:	PWR	PWR
Nuclear Steam Supply System Vendor:	WEST	WEST

Cooling Water System

Type: Once-through and/or natural draft cooling towers
Source: Chickamauga Lake
Source Temperature Range: 42–83°F (6–28°C)
Condenser Flow Rate: 522,000 gpm (32.9 m^3/s) each unit
Design Condenser Temperature Rise: 30°F (17°C)
Intake Structure: Intake from lake
Discharge Structure: Discharge to lake

Site Information

Total Area: 525 ac (212 ha)
Exclusion Distance: 0.35 mi (0.56 km)
Low Population Zone: 3 mi (4.83 km)
Nearest City: Chattanooga; 2000 population: 155,554
Site Topography: Rolling
Surrounding Area Topography: Hilly
Dominant Land Cover within 5 mi (8 km): Forest, agriculture, open water
Level 1 Ecoregion within 5 mi (8 km): Eastern Temperate Forest
Level 3 Ecoregion within 5 mi (8 km): Ridge and Valley
Percent Wetland within 5 mi (8 km): 15.1, mostly lake

Nearby Features: The nearest town is Shady Grove about 2 mi (3 km) NW. Harrison Bay State Park is 3 mi (5 km) S. The Volunteer Ordnance Works is about 9 mi (15 km) S. Chickamauga Lake is part of the Tennessee River.

Population within a 50 mi (80 km) Radius: 954,430

SHEARON HARRIS NUCLEAR POWER PLANT

Location: Wake County, North Carolina
20 mi (32 km) SW of Raleigh
Latitude 35.6336°N; longitude 78.9564°W
Licensee: Progress Energy

Unit Information	Unit 1
Docket Number:	50-400
Construction Permit:	1978
Operating License:	1987
Commercial Operation:	1987
License Expiration:	2046
Licensed Thermal Power (MWt):	2,900
Net Capacity (MWe):	900
Type of Reactor:	PWR
Nuclear Steam Supply System Vendor:	WEST

Cooling Water System

Type: Natural draft cooling tower
Source: Buckhorn Creek
Source Temperature Range: 41–81°F (5–27°C)
Condenser Flow Rate: 483,000 gpm (30.5 m^3/s)
Design Condenser Temperature Rise: 25.7°F (14.3°C)
Intake Structure: At shoreline of reservoir on Buckhorn Creek
Discharge Structure: Discharged to reservoir

Site Information

Total Area: 10,744 ac (4,348 ha)
Exclusion Distance: 6,640 ft (2 km) (northwest) to 7,000 ft (2.1 km) (east) to 7,200 ft (2.2 km) (south)
Low Population Zone: 3 mi (4.83 km)
Nearest City: Raleigh; 2000 population: 276,093
Site Topography: Rolling
Surrounding Area Topography: Rolling
Dominant Land Cover within 5 mi (8 km): Forest, herbaceous, open water
Level 1 Ecoregion within 5 mi (8 km): Eastern Temperate Forest
Level 3 Ecoregion within 5 mi (8 km): Piedmont; Southeastern Plains

Percent Wetland within 5 mi (8 km): 13.2, mostly lake
Nearby Features: The nearest town is Bonsal 2 mi (3 km) NW. The Seaboard Coast Line Railroad is 2 mi (3 km) NW. Buckhorn Creek feeds into the Cape Fear River.
Population within a 50 mi (80 km) Radius: 2,035,797

SOUTH TEXAS PROJECT NUCLEAR GENERATING STATION

Location:	Matagorda County, Texas 12 mi (19 km) SSW of Bay City Latitude 28.7950°N; longitude 96.0481°W
Licensee:	STP Nuclear Operating Co.

Unit Information	Unit 1	Unit 2
Docket Number:	50-498	50-499
Construction Permit:	1975	1975
Operating License:	1988	1989
Commercial Operation:	1988	1989
License Expiration:	2027	2028
Licensed Thermal Power (MWt):	3,853	3,853
Net Capacity (MWe):	1,280	1,280
Type of Reactor:	PWR	PWR
Nuclear Steam Supply System Vendor:	WEST	WEST

Cooling Water System

Type: Closed cycle cooling reservoir
Source: Colorado River
Source Temperature Range: 58–84°F (14–29°C)
Condenser Flow Rate: 907,400 gpm (57.26 m^3/s) each unit
Design Condenser Temperature Rise: 19°F (11°C)
Intake Structure: On bank of Colorado River
Discharge Structure: On bank of Colorado River

Site Information

Total Area: 12,350 ac (4,998 ha)
Exclusion Distance: 0.89 mi (1.43 km) minimum
Low Population Zone: 3 mi (4.83 km)
Nearest City: Galveston; 2000 population: 57,247
Site Topography: Flat
Surrounding Area Topography: Flat
Dominant Land Cover within 5 mi (8 km): Agriculture, open water, wetland
Level 1 Ecoregion within 5 mi (8 km): Great Plains
Level 3 Ecoregion within 5 mi (8 km): Western Gulf Coastal Plain
Percent Wetland within 5 mi (8 km): 23.3, mostly lake

Nearby Features: The nearest town is Matagorda 8 mi (13 km) SE. The Missouri Pacific Railroad is about 5 mi (8 km) NNE. A 16-in. (40-cm) natural gas pipeline is about 2 mi (3 km) NW.

Population within a 50 mi (80 km) Radius: 402,902

SURRY POWER STATION

Location:	Surry County, Virginia 17 mi (27 km) NW of Newport News Latitude 37.1656°N; longitude 76.6983°W
Licensee:	Dominion Generation

Unit Information	Unit 1	Unit 2
Docket Number:	50-280	20-281
Construction Permit:	1968	1968
Operating License:	1972	1973
Commercial Operation:	1972	1973
License Expiration:	2032	2033
Licensed Thermal Power (MWt):	2,546	2,546
Net Capacity (MWe):	799	799
Type of Reactor:	PWR	PWR
Nuclear Steam Supply System Vendor:	WEST	WEST

Cooling Water System

Type: Once-through
Source: James River
Source Temperature Range: 35–84°F (2–29°C)
Condenser Flow Rate: 1.68 million gpm (106 m^3/s) both units
Design Condenser Temperature Rise: 14°F (7.8°C)
Intake Structure: 1.7 mi (2.7 km) concrete canal
Discharge Structure: 2,900 ft (880 m) canal

Site Information

Total Area: 840 ac (340 ha)
Exclusion Distance: 1,650 ft (500 m) radius or 0.31 mi (0.5 km)
Low Population Zone: 3 mi (4.83 km)
Nearest City: Newport News; 2000 population: 180,150
Site Topography: Flat
Surrounding Area Topography: Flat
Dominant Land Cover within 5 mi (8 km): Open water, forest, agriculture
Level 1 Ecoregion within 5 mi (8 km): Eastern Temperate Forest
Level 3 Ecoregion within 5 mi (8 km): Middle Atlantic Coastal Plain; Southeastern Plains
Percent Wetland within 5 mi (8 km): 60.9, mostly estuarine and marine deepwater; riverine

Nearby Features: The nearest town is Scotland 5 mi (8 km) W. Jamestown Island, a Federal park, is 4 mi (6 km) NW. Chippokes Plantation, a State park, is 3 mi (5 km) WSW. Jamestown National Historical Park is 5 mi (8 km) WNW. Colonial Williamsburg is 7 mi (11 km) NNW. Adjacent to the site on the north is Hog Island, a waterfowl refuge. U.S. Highway I-64 is 12 mi (19 km) NW.

Population within a 50 mi (80 km) Radius: 2,387,353

SUSQUEHANNA STEAM ELECTRIC STATION

Location:	Luzerne County, Pennsylvania 7 mi (11 km) NE of Berwick Latitude 41.0922°N; longitude 76.1467°W
Licensee:	PPL Susquehanna, LLC

Unit Information	Unit 1	Unit 2
Docket Number:	50-387	50-388
Construction Permit:	1973	1973
Operating License:	1982	1984
Commercial Operation:	1983	1985
License Expiration:	2042	2044
Licensed Thermal Power (MWt):	3,952	3,952
Net Capacity (MWe):	1,149	1,140
Type of Reactor:	BWR	BWR
Nuclear Steam Supply System Vendor:	GE	GE

Cooling Water System

Type: Natural draft cooling towers
Source: Susquehanna River
Source Temperature Range: Not available
Condenser Flow Rate: 968,000 gpm (61 m^3/s) both units
Design Condenser Temperature Rise: 14°F (8°C)
Intake Structure: Intake bays on river bank
Discharge Structure: Diffuser pipe 200 ft (61 m) from river bank

Site Information

Total Area: 1,173 ac (475 ha)
Exclusion Distance: 0.34 mi (0.55 km) radius
Low Population Zone: 3 mi (4.83 km)
Nearest City: Wilkes-Barre; 2000 population: 43,123
Site Topography: Rolling
Surrounding Area Topography: Hilly with flat river valley
Dominant Land Cover within 5 mi (8 km): Forest, agriculture, developed: open space
Level 1 Ecoregion within 5 mi (8 km): Eastern Temperate Forest
Level 3 Ecoregion within 5 mi (8 km): Ridge and Valley
Percent Wetland within 5 mi (8 km): 4.6, mostly riverine

Nearby Features: The nearest town is Beach Haven about 1 mi (1.6 km) SW. U.S. Highway I-80 is 5 mi (8 km) E, and the Delaware and Hudson Railroad is 1 mi (1.6 km) E.

Population within a 50 mi (80 km) Radius: 1,684,794

THREE MILE ISLAND, UNIT 1

Location: Dauphin County, Pennsylvania
10 mi (16 km) SE of Harrisburg
Latitude 40.1531°N; longitude 76.7250°W
Licensee: Exelon Generation Company

Unit Information	Unit 1
Docket Number:	50-289
Construction Permit:	1968
Operating License:	1974
Commercial Operation:	1974
License Expiration:	2034
Licensed Thermal Power (MWt):	2,568
Net Capacity (MWe):	786
Type of Reactor:	PWR
Nuclear Steam Supply System Vendor:	B&W

Cooling Water System

Type: Natural draft cooling towers
Source: Susquehanna River
Source Temperature Range: 33–84°F (1–29°C)
Condenser Flow Rate: 430,000 gpm (27 m^3/s)
Design Condenser Temperature Rise: Not available
Intake Structure: Concrete structure on river bank
Discharge Structure: Discharged at the shoreline

Site Information

Total Area: 472 ac (191 ha)
Exclusion Distance: 0.38 mi (0.61 km) radius
Low Population Zone: 2 mi (3.22 km)
Nearest City: Harrisburg; 2000 population: 48,950
Site Topography: Flat
Surrounding Area Topography: Rolling to hilly
Dominant Land Cover within 5 mi (8 km): Agriculture, forest, developed: high, medium, low density
Level 1 Ecoregion within 5 mi (8 km): Eastern Temperate Forest
Level 3 Ecoregion within 5 mi (8 km): Northern Piedmont

Percent Wetland within 5 mi (8 km): 11.4, mostly riverine
Nearby Features: The nearest town is Middletown 4 mi (6 km) N. Harrisburg-York Airport is 8 mi (13 km) WNW. Unit 2 ceased operation after an accident in 1979. Peach Bottom Atomic Power Station is 35 mi (56 km) downstream.
Population within a 50 mi (80 km) Radius: 2,466,679

TURKEY POINT NUCLEAR PLANT

Location:	Dade County, Florida 25 mi (40 km) S of Miami Latitude 25.4350°N; longitude 80.3314°W
Licensee:	Florida Power and Light Co.

Unit Information	Unit 3	Unit 4
Docket Number:	50-250	50-251
Construction Permit:	1967	1967
Operating License:	1972	1973
Commercial Operation:	1972	1973
License Expiration:	2032	2033
Licensed Thermal Power (MWt):	2,300	2,300
Net Capacity (MWe):	693	693
Type of Reactor:	PWR	PWR
Nuclear Steam Supply System Vendor:	WEST	WEST

Cooling Water System

Type: Closed-cycle cooling canal
Source: Biscayne Bay
Source Temperature Range: 54–90°F (12–32°C)
Condenser Flow Rate: 1.3 million gpm (82 m^3/s) both units
Design Condenser Temperature Rise: 18°F (10°C)
Intake Structure: Intake canal and barge canal
Discharge Structure: Canal system covering about 4,000 ac (1,600 ha)

Site Information

Total Area: 24,000 ac (9,700 ha)
Exclusion Distance: 0.79 mi (1.27 km)
Low Population Zone: 5 mi (8.05 km)
Nearest City: Miami; 2000 population: 362,470
Site Topography: Flat
Surrounding Area Topography: Flat
Dominant Land Cover within 5 mi (8 km): Wetland, open water, agriculture
Level 1 Ecoregion within 5 mi (8 km): Tropical Wet Forest
Level 3 Ecoregion within 5 mi (8 km): Southern Florida Coastal Plain
Percent Wetland within 5 mi (8 km): 91.5, mostly estuarine and marine deepwater

Nearby Features: The nearest town is Florida City about 9 mi (14 km) W. Hawk Missile Base is 1 mi (1.6 km) NW. Homestead Recreation Park is about 2 mi (3 km) NNW. The Florida East Coast Railroad is about 9 mi (14 km) NW. Units 1 and 2 are coal-fired and adjacent to the site.

Population within a 50 mi (80 km) Radius: 7,490,123

VERMONT YANKEE NUCLEAR POWER STATION

Location: Windham County, Vermont
5 mi (8 km) S of Brattleboro
Latitude 42.7803°N; longitude 72.5158°W
Licensee: Entergy Nuclear Operations, Inc.

Unit Information	Unit 1
Docket Number:	50-271
Construction Permit:	1967
Operating License:	1973
Commercial Operation:	1972
License Expiration:	2032
Licensed Thermal Power (MWt):	1,912
Net Capacity (MWe):	510
Type of Reactor:	BWR
Nuclear Steam Supply System Vendor:	GE

Cooling Water System

Type: Once-through; closed-cycle mechanical draft towers
Source: Connecticut River
Source Temperature Range: 0–23°F (32–74°C)
Condenser Flow Rate: 360,000 gpm (22.7 m^3/s)
Design Condenser Temperature Rise: 13.4°F (10°C)
Intake Structure: Concrete structure at edge of river
Discharge Structure: Aerating structure discharges at edge of river

Site Information

Total Area: 125 ac (50.6 ha)
Exclusion Distance: 0.17 mi (0.27 km)
Low Population Zone: 5 mi (8.05 km)
Nearest City: Holyoke, Massachusetts; 2000 population: 39,838
Site Topography: Flat
Surrounding Area Topography: Rolling to hilly
Dominant Land Cover within 5 mi (8 km): Forest, agriculture, developed: high, medium, low density
Level 1 Ecoregion within 5 mi (8 km): Eastern Temperate Forest
Level 3 Ecoregion within 5 mi (8 km): Northeastern Coastal Zone; Northeastern Highlands

Percent Wetland within 5 mi (8 km): 6.1, mostly lake
Nearby Features: The nearest town is Vernon about 1 mi (1.6 km) W. Vernon Dam is 0.7 mi (1 km) downstream from the site. The decommissioned Yankee Nuclear Power Station is about 20 mi (32 km) WSW.
Population within a 50 mi (80 km) Radius: 1,513,282

VIRGIL C. SUMMER NUCLEAR STATION

Location:	Fairfield County, South Carolina 26 mi (42 km) NW of Columbia Latitude 34.2958°N; longitude 81.3203°W
Licensee:	South Carolina Electric & Gas Co.

Unit Information	Unit 1
Docket Number:	50-395
Construction Permit:	1973
Operating License:	1982
Commercial Operation:	1984
License Expiration:	2042
Licensed Thermal Power (MWt):	2,900
Net Capacity (MWe):	966
Type of Reactor:	PWR
Nuclear Steam Supply System Vendor:	WEST

Cooling Water System

Type: Once-through
Source: Lake Monticello
Source Temperature Range: 52–91°F (11–33°C)
Condenser Flow Rate: 507,000 gpm (32 m^3/s)
Design Condenser Temperature Rise: 25°F (14°C)
Intake Structure: Intake at shoreline
Discharge Structure: Discharge to lake via a discharge basin and 1,000-ft (305-m) canal

Site Information

Total Area: 2,200 ac (890 ha)
Exclusion Distance: 1.01 mi (1.63 m) radius
Low Population Zone: 3 mi (4.83 km)
Nearest City: Columbia; 2000 population: 116,278
Site Topography: Rolling
Surrounding Area Topography: Rolling to hilly
Dominant Land Cover within 5 mi (8 km): Forest, open water, herbaceous
Level 1 Ecoregion within 5 mi (8 km): Eastern Temperate Forest
Level 3 Ecoregion within 5 mi (8 km): Piedmont
Percent Wetland within 5 mi (8 km): 20.2, mostly lake

Nearby Features: The nearest town is Jenkinsville 3 mi (5 km) SE. U.S. Highway I-26 is 7 mi (11 km) SSW. The Southern Railroad is 1 mi (1.6 km) W. The Fairfield pumped storage hydrostation is about 1 mi (1.6 km) NW and uses Lake Monticello as well as the Parr Reservoir.

Population within a 50 mi (80 km) Radius: 1,032,330

VOGTLE ELECTRIC GENERATING PLANT

Location: Burke County, Georgia
26 mi (42 km) SE of Augusta
Latitude 33.1414°N; longitude 81.7625°W
Licensee: Southern Nuclear Operating Co.

Unit Information	Unit 1	Unit 2
Docket Number:	50-424	50-425
Construction Permit:	1974	1974
Operating License:	1987	1989
Commercial Operation:	1987	1989
License Expiration:	2047	2049
Licensed Thermal Power (MWt):	3,565	3,565
Net Capacity (MWe):	1,109	1,127
Type of Reactor:	PWR	PWR
Nuclear Steam Supply System Vendor:	WEST	WEST

Cooling Water System

Type: Natural draft cooling towers
Source: Savannah River
Source Temperature Range: 39–86°F (4–30°C)
Condenser Flow Rate: 509,600 gpm (32.16 m^3/s) each unit
Design Condenser Temperature Rise: 33°F (18°C)
Intake Structure: At river bank
Discharge Structure: Single-point discharge pipe near the shoreline

Site Information

Total Area: 3,169 ac (1,282 ha)
Exclusion Distance: 0.68 mi (1.09 km) minimum
Low Population Zone: 2 mi (3.22 km) radius
Nearest City: Augusta-Richmond County; 2000 population: 195,182
Site Topography: Rolling
Surrounding Area Topography: Rolling, river flood plain
Dominant Land Cover within 5 mi (8 km): Forest, wetland, herbaceous
Level 1 Ecoregion within 5 mi (8 km): Eastern Temperate Forest
Level 3 Ecoregion within 5 mi (8 km): Southeastern Plains
Percent Wetland within 5 mi (8 km): 27.4, mostly freshwater forested/shrub wetland

Nearby Features: The nearest town is Shell Bluff about 7 mi (11 km) W. The Seaboard Coast Line Railroad is about 4 mi (6 km) NE. The Department of Energy Savannah River Plant is about 10 mi (16 km) NNE.

Population within 50 mi (80 km) Radius: 670,000

WATERFORD STEAM ELECTRIC STATION

Location:	St. Charles County, Louisiana 20 mi (32 km) W of New Orleans Latitude 29.9947°N; longitude 90.4711°W
Licensee:	Entergy Nuclear Operations, Inc.

Unit Information	Unit 3
Docket Number:	50-382
Construction Permit:	1974
Operating License:	1985
Commercial Operation:	1985
License Expiration:	2024
Licensed Thermal Power (MWt):	3,716
Net Capacity (MWe):	1,250
Type of Reactor:	PWR
Nuclear Steam Supply System Vendor:	CE

Cooling Water System

Type: Once-through
Source: Mississippi River
Source Temperature Range: 46–82°F (8–28°C)
Condenser Flow Rate: 975,000 gpm (61.53 m^3/s)
Design Condenser Temperature Rise: 16°F (9°C)
Intake Structure: At river bank
Discharge Structure: At river bank

Site Information

Total Area: 3,561 ac (1,441 ha)
Exclusion Distance: 90.57 mi (0.92 km) radius
Low Population Zone: 2 mi (3.22 km)
Nearest City: New Orleans; 2000 population: 484,674
Site Topography: Flat
Surrounding Area Topography: Flat
Dominant Land Cover within 5 mi (8 km): Wetland, agriculture, developed: high, medium, low density
Level 1 Ecoregion within 5 mi (8 km): Eastern Temperate Forest
Level 3 Ecoregion within 5 mi (8 km): Mississippi Alluvial Plain

Percent Wetland within 5 mi (8 km): 67.8, mostly freshwater forested/shrub wetland
Nearby Features: The nearest town is Killona 1 mi (1.6 km) WNW. U.S. Highway I-10 is about 7 mi (11 km) NE and I-90 about 7 mi (11 km) SE. Several active and abandoned gas and oil fields are within 10 mi (16 km). Lake Pontchartrain is about 7 mi (11 km) NE. The Missouri Pacific Railroad is just S of the site, and the Southern Pacific Railroad is about 8 mi (13 km) SE.
Population within a 50 mi (80 km) Radius: 2,072,270

WATTS BAR NUCLEAR PLANT

Location: Rhea County, Tennessee
7 mi (11 km) SSE of Spring City
Latitude 35.6022°N; longitude 84.7894°W
Licensee: Tennessee Valley Authority

Unit Information	Unit 1[(a)]
Docket Number:	50-390
Construction Permit:	1973
Operating License:	1996
Commercial Operation:	1996
License Expiration:	2035
Licensed Thermal Power (MWt):	3,459
Net Capacity (MWe):	1,123
Type of Reactor:	PWR
Nuclear Steam Supply System Vendor:	WEST

Cooling Water System

Type: Natural draft cooling towers
Source: Chickamauga Lake
Source Temperature Range: 43–82°F (6–28°C)
Condenser Flow Rate: 410,000 gpm (26 m^3/s) each unit
Design Condenser Temperature Rise: 38°F (21°C)
Intake Structure: At lake bank
Discharge Structure: To lake via a holding pond

Site Information

Total Area: 1,770 ac (716 ha)
Exclusion Distance: 0.75 mi (1.21 km) radius
Low Population Zone: 3 mi (4.83 km)
Nearest City: Chattanooga; 2000 population: 155,554
Site Topography: Flat to rolling
Surrounding Area Topography: Rolling to hilly
Dominant Land Cover within 5 mi (8 km): Forest, agriculture, open water

---

(a) Construction of Unit 1 was halted in 1985, resumed in 1990, and completed in 1995. Construction of Unit 2 was halted in 1988. The unit remains idle.

Level 1 Ecoregion within 5 mi (8 km): Eastern Temperate Forest
Level 3 Ecoregion within 5 mi (8 km): Ridge and Valley
Percent Wetland within 5 mi (8 km): 10.9, mostly lake
Nearby Features: The nearest town is Peakland 2 mi (3 km) NE. Watts Bar Dam is 1 mi (1.6 km) N. A fossil-fired steam plant is just N of the site. U.S. Highway I-75 is about 11 mi (18 km) SE. The New Orleans and Texas Pacific Railroad is 7 mi (11 km) NW. Chickamauga Lake is on the Tennessee River.
Population within a 50 mi (80 km) Radius: 1,044,454

WOLF CREEK GENERATING STATION

Location:	Coffey County, Kansas 4 mi (6 km) NE of Burlington Latitude 38.2386°N; longitude 95.6894°W
Licensee:	Wolf Creek Nuclear Operating Corporation

Unit Information	Unit 1
Docket Number:	50-482
Construction Permit:	1977
Operating License:	1985
Commercial Operation:	1985
License Expiration:	2045
Licensed Thermal Power (MWt):	3,565
Net Capacity (MWe):	1,166
Type of Reactor:	PWR
Nuclear Steam Supply System Vendor:	WEST

Cooling Water System

Type: Closed-cycle cooling pond
Source: Wolf Creek
Source Temperature Range: 32–87°F (0–31°C)
Condenser Flow Rate: 500,000 gpm (30 m^3/s)
Design Condenser Temperature Rise: 30°F (1.1°C)
Intake Structure: On the shore of cooling lake
Discharge Structure: Discharged to 5,090 ac (2,060 ha) cooling lake, into an embayment separated from the intake

Site Information

Total Area: 9,818 ac (3,973 ha)
Exclusion Distance: 0.75 mi (1.21 km) radius
Low Population Zone: 2.5 mi (4.02 km) radius
Nearest City: Topeka; 2000 population: 122,377
Site Topography: Flat to rolling
Surrounding Area Topography: Flat to rolling
Dominant Land Cover within 5 mi (8 km): Herbaceous, agriculture, open water
Level 1 Ecoregion within 5 mi (8 km): Great Plains
Level 3 Ecoregion within 5 mi (8 km): Central Irregular Plains

Percent Wetland within 5 mi (8 km): 14.8, mostly lake

Nearby Features: The nearest town is Sharpe about 2 mi (3 km) N. The Flint Hills National Wildlife Refuge is about 7 mi (11 km) W. The John Redmond Reservoir is about 4 mi (6 km) W. U.S. Highway I-35 is 14 mi (23 km) N. The cooling lake is formed by a dam on Wolf Creek.

Population within a 50 mi (80 km) Radius: 176,301

References

Commission for Environmental Cooperation (CEC). 2006. *Level I Ecological Regions of North America, Commission for Environmental Cooperation*. Available URL: http://www.epa.gov/wed/pages/ecoregions/na_eco.htm#Level%20I (Accessed November 6, 2007).

Energy Information Administration (DOE/EIA). 2007a. *Nuclear Power Plants Operating in the United States as of September 30, 2005*. Available URL: http://www.eia.doe.gov/cneaf/nuclear/page/at_a_glance/reactors/nuke1.html (Accessed December 5, 2007).

Energy Information Administration (DOE/EIA). 2007b. *Monthly Nuclear Utility Generation by State and Reactor, 2007—January through October*. Available URL: http://www.eia.doe.gov/cneaf/nuclear/page/nuc_generation/usreactors.xls (Accessed February 6, 2008).

U.S. Census Bureau (USCB). 2007. *State and County Quickfacts*. Available URL: http://quickfacts.census.gov/qfd/index.html (Accessed December 5, 2007).

U.S. Environmental Protection Agency (EPA). 2007. *Level III Ecoregions of the Conterminous United States*. National and Environmental Effects Research Laboratory. Revised March, 2007. Available URL: http://www.epa.gov/wed/pages/ecoregions/level_iii.htm (Accessed November 6, 2007).

U.S. Geological Survey (USGS). 2003. *National Land Cover Database*. Sioux Falls, South Dakota.

U.S. Fish and Wildlife Service (USFWS). 2007. *National Wetlands Inventory*. Available URL: http://wetlandsfws.er.usgs.gov/NWI/index.html (Accessed November 7, 2007).

U.S. Nuclear Regulatory Commission (NRC). 1996. *Generic Environmental Impact Statement for License Renewal of Nuclear Plants*. NUREG-1437, Vols. 1 and 2. Washington, D.C.

U.S. Nuclear Regulatory Commission (NRC). 2008a. *Approved Applications for Power Uprates*. Available URL: http://www.nrc.gov/reactors/operation/licensing/power-uprates/approved-applications.html (Accessed February 6, 2008).

U.S. Nuclear Regulatory Commission (NRC). 2008b. *Power Uprates for Nuclear Plants*. Available URL: http://www.nrc.gov/reading-rm/doc-collections/fact-sheets/power-uprates.html (Accessed October 31, 2010).

U.S. Nuclear Regulatory Commission (NRC). 2010. *List of Power Reactor Units*. Available URL: http://www.nrc.gov/reactors/operating/list-power-reactor-units.html (Accessed October 5, 2010).

Appendix D

Technical Support for GEIS Analyses

Appendix D

Technical Support for GEIS Analyses

D.1 Land Use and Visual Resources

D.1.1 Description of Affected Resources and Region of Influence

Onsite land use resources that could be affected by continued power plant operations during the license renewal term include all of the land within the plant site property boundaries. For license renewal, the current onsite industrial use of the land use is assumed to remain unchanged. Offsite land use resources include all patterns of land use in the vicinity of the nuclear plant that could be affected by continued operations and refurbishment activities associated with license renewal. Transmission lines generally have no effect on land use within the right-of-way (ROW) corridor. The region of influence for visual resource impacts includes all areas within view of the nuclear power plant.

D.1.2 Description of Impact Assessment

Completed license renewal supplemental environmental impact statements (SEISs) were examined to determine the extent of past onsite land use disturbances occurring as a result of license renewal and refurbishment activities at nuclear power plants. Offsite land use impacts were assessed based on a survey of newspaper and magazine accounts related to land use issues, a survey of all operating nuclear power plants, and experiences at seven selected license renewal case study sites. Additional information on offsite land use impacts came from a literature search on land use controls and workforce impacts with respect to nuclear power plants. The evaluation of land use impacts of transmission lines was derived from information compiled from a review of license renewal SEISs. The assessment of visual resource impacts caused by the presence of the nuclear power plant and transmission lines was derived from a review of license renewal SEISs.

D.2 Air Quality and Noise

D.2.1 Description of Affected Resources and Region of Influence

Similar to most industrial facilities, nuclear power plants and other fuel-cycle facilities generate air pollutants[(a)] and propagate noise. The region of influence of the effects on air quality and noise includes the regulated ambient air environment within and outside each plant site property boundaries, whether open to the public or under various levels of access control. For license renewal, current air and noise pollution levels would remain unchanged, while during refurbishment and decommissioning activities, some additional impacts could occur, particularly on plant property. Some offsite areas at a few plants have a limited region of influence of a few kilometers beyond plant boundaries for measurable air quality impacts and a few hundred meters for noticeable noise impacts. The region of air and noise influences for primary source activities can conservatively include downwind areas within a 15.5 to 31 mi (25 to 50 km) radius of the plant for air impacts and up to about 1.9 mi (3 km) from the fence line for noise impacts. It is not anticipated that construction or refurbishment activities associated with license renewal would involve new primary sources of air or noise pollution.

D.2.2 Description of Impact Assessment

Air quality and noise impacts were examined through a review of a select number of SEISs to identify air emission sources that are permitted or have applications for permits during license renewal, as well as the information provided in the 1996 GEIS (NRC 1996) and the decommissioning GEIS (NRC 2002), which addressed impacts for plants that have undergone decommissioning or are in the process of decommissioning. Offsite impacts were assessed based on a survey of all operating nuclear power plants and information in the decommissioning GEIS.

The following, including figures and tables, provides supplemental data and information in support of the air quality impacts provided in Sections 3.3.1, 3.3.2, and 4.3.1.1.

D.2.2.1 Climatology

Continental U.S. maximum and minimum average annual temperatures from 1981 through 2010 are shown in Figures D.2-1 and D.2-2, respectively. The average annual precipitation over the same period is shown in Figure D.2-3. In the period from 2006 through 2010, actual precipitation as a percent of the average monthly precipitation is shown in Figure D.2-4.

(a) Both radiological and nonradiological (criteria air pollutants) releases are covered in the GEIS. See Appendix F for description of region of influence and the impact assessment for radiological releases.

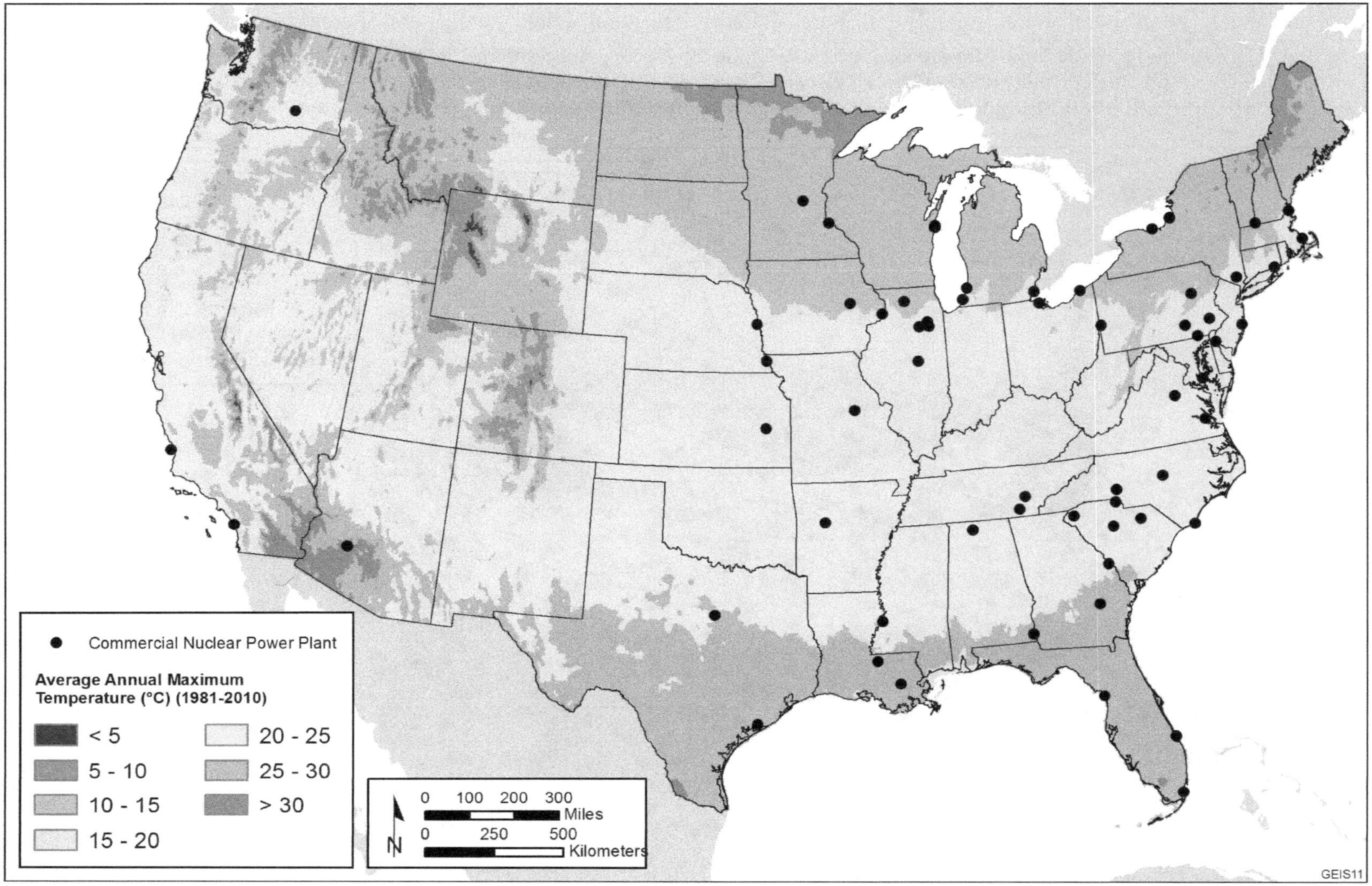

Figure D.2-1. Average Annual Maximum Temperatures over the Continental United States (1981–2010)
(Permission to use this copyrighted material is granted by PRISM Group, Oregon State University)
Copyright © 2010, PRISM Climate Group, Oregon State University, http://www.prismclimate.org. Map created February 17, 2012.

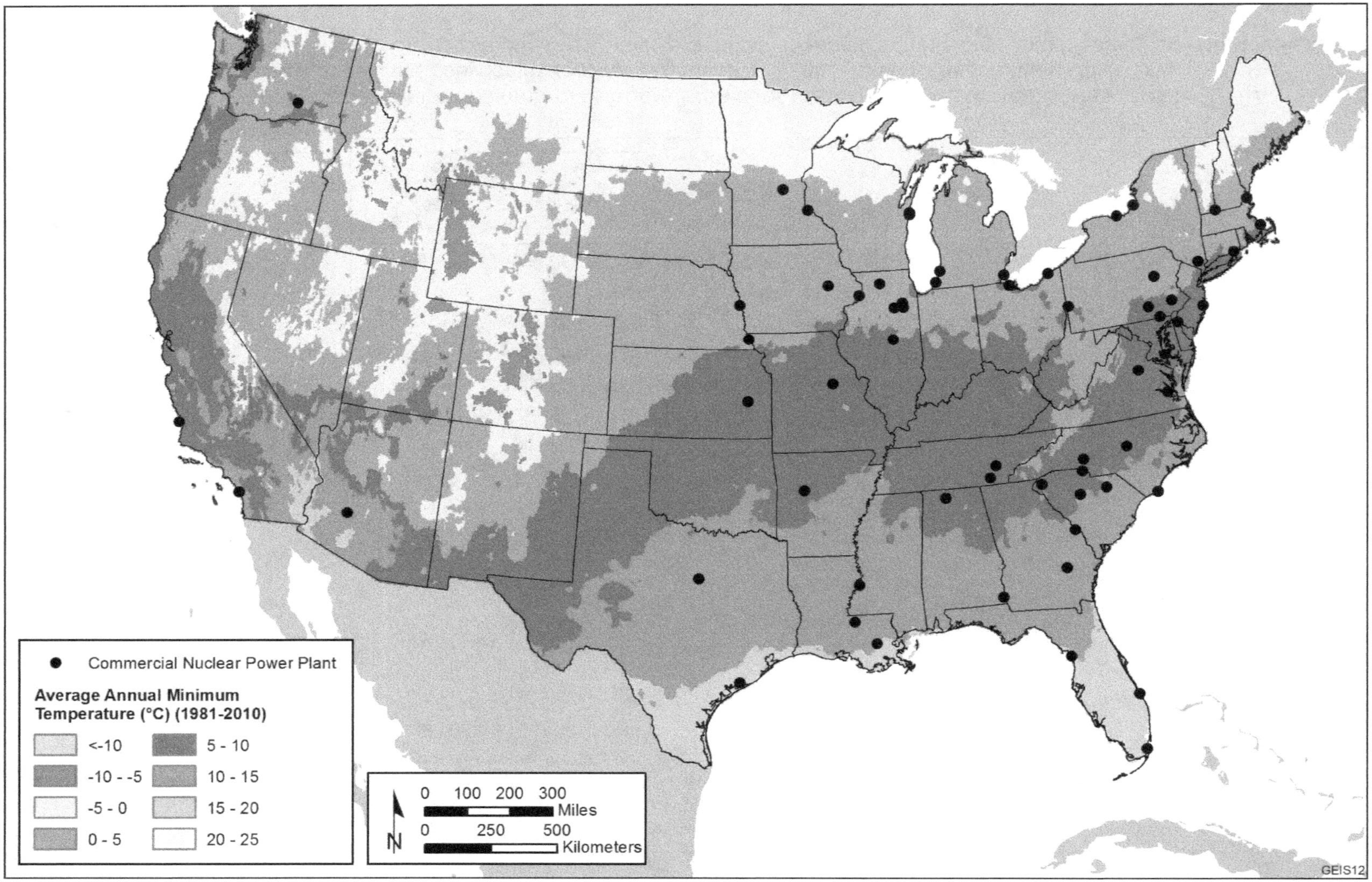

Figure D.2-2. Average Annual Minimum Temperatures over the Continental United States (1981–2010)
(Permission to use this copyrighted material is granted by PRISM Group, Oregon State University)
Copyright © 2010, PRISM Climate Group, Oregon State University, http://www.prismclimate.org. Map created February 17, 2012.

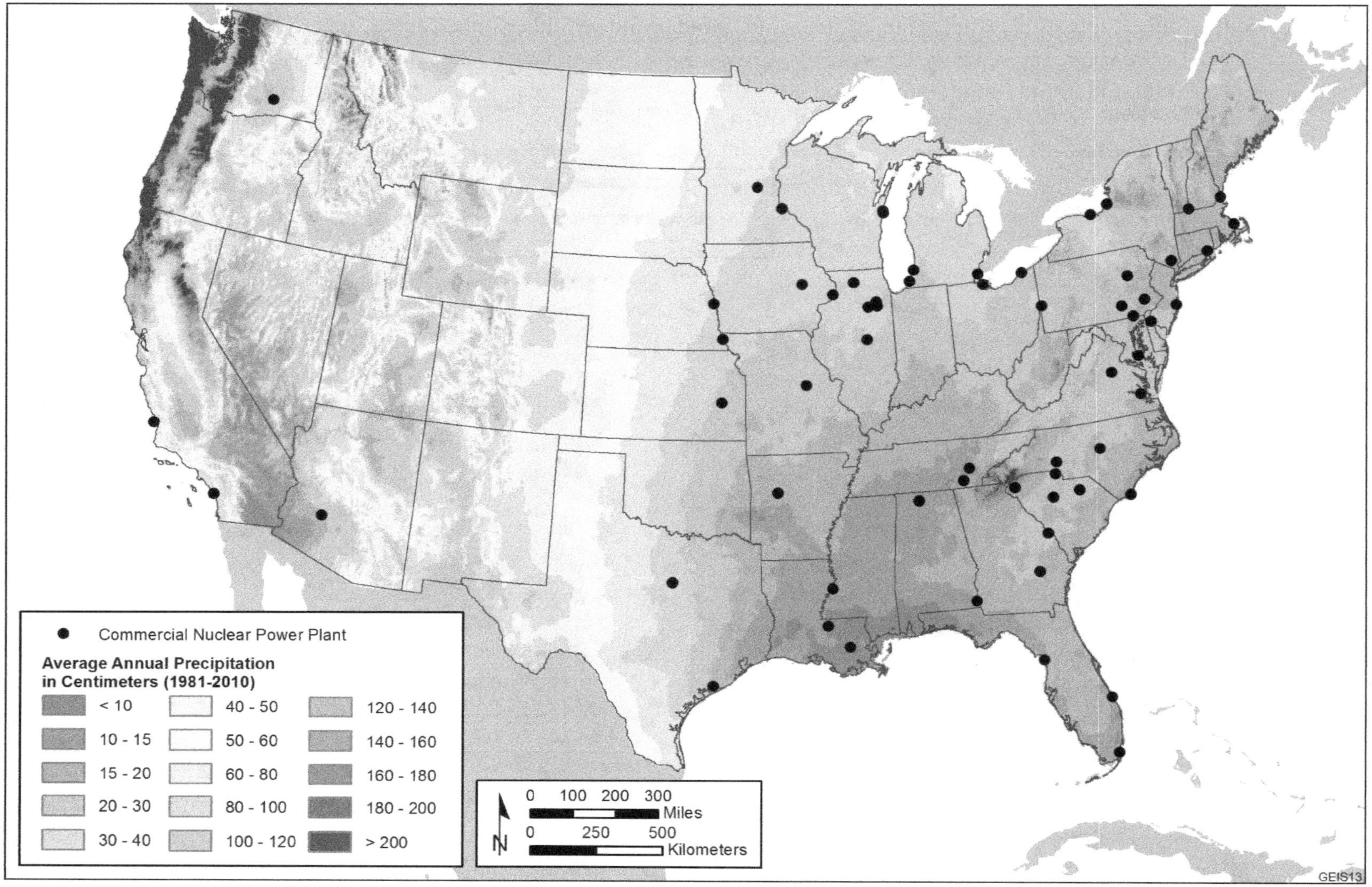

Figure D.2-3. Average Annual Precipitation over the Continental United States (1981–2010)
(Permission to use this copyrighted material is granted by PRISM Group, Oregon State University)
Copyright © 2010, PRISM Climate Group, Oregon State University, http://www.prismclimate.org. Map created February 17, 2012.

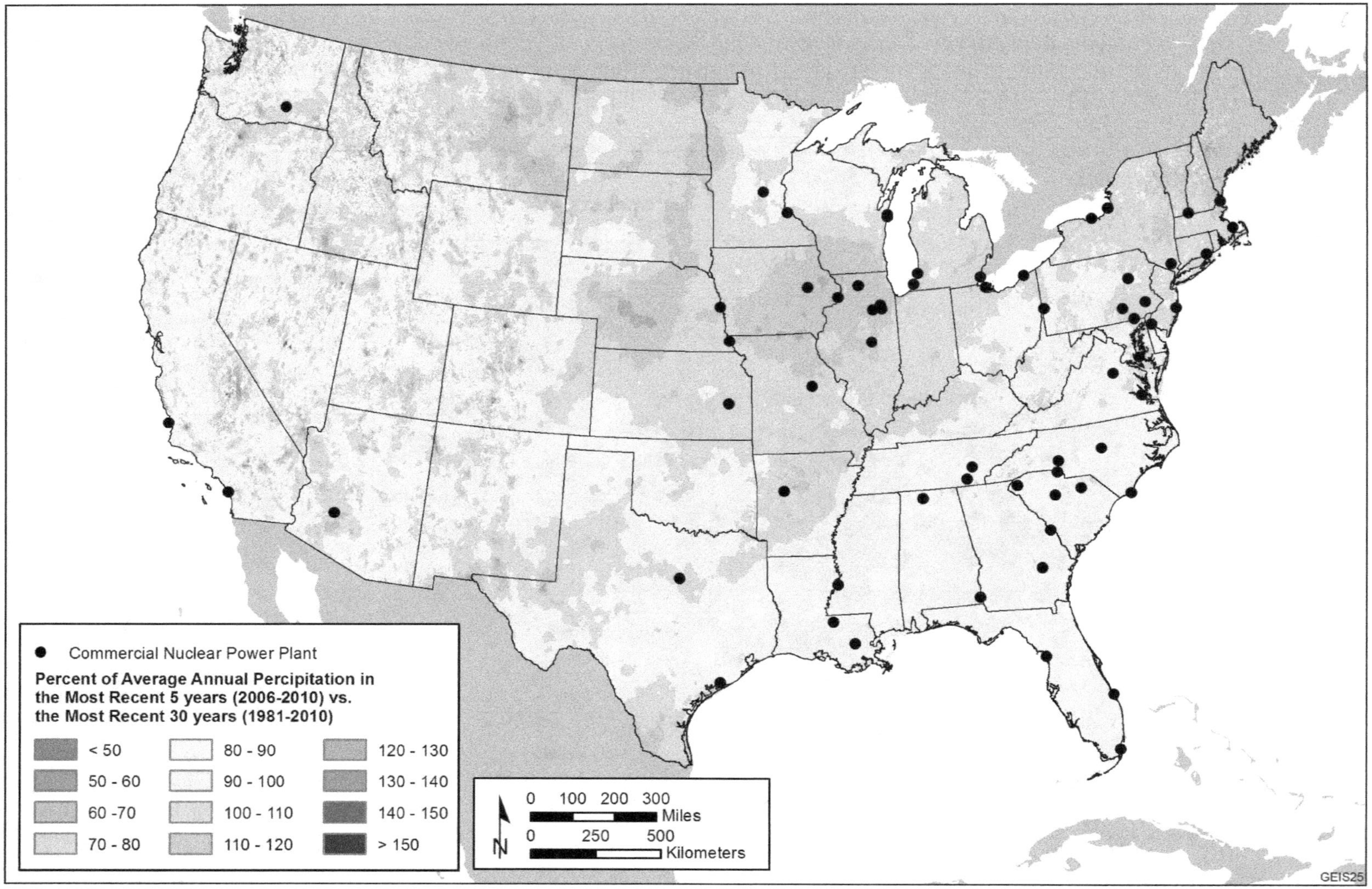

Figure D.2-4. Percent of Average Monthly Precipitation over the Past 5 Years (2006–2010) vs. the Past 30 Years (1981–2010)

(Permission to use this copyrighted material is granted by PRISM Group, Oregon State University)

Copyright © 2010, PRISM Climate Group, Oregon State University, http://www.prismclimate.org. Map created February 17, 2012.

Drought or near-drought conditions are shown in south-central California northeast of the San Onofre plant and north of the Diablo Canyon plant. Similar drought conditions over this recent 5-year period also appear in limited areas near the Palo Verde plant in Arizona. Above normal annual precipitation (10 to 30 percent above historical averages) in the vicinity of licensed commercial power reactors are shown in large areas of southwestern and south Texas, the Midwest, and over much of the northeastern United States.

D.2.2.2 Air Quality

Air quality in all geographical regions of the United States is classified as being either in attainment or nonattainment. The U.S. Environmental Protection Agency (EPA) has the authority to formally designate areas as attainment or nonattainment areas and uses the National Ambient Air Quality Standards (NAAQS) to evaluate an area's attainment status. The pollutants for which the NAAQS have been established are known as criteria pollutants (ozone [O_3], particulate matter with an aerodynamic diameter of less than or equal to 10 μm [PM_{10}], particulate matter with an aerodynamic diameter of less than or equal to 2.5 μm [$PM_{2.5}$], nitrogen dioxide [NO_2], SO_2, CO, and lead [Pb]). If the concentration limit of a pollutant is below the NAAQS, the area will be designated as being in attainment for that pollutant. An area is deemed to be in attainment by the EPA when the air quality is monitored and the resultant concentrations are found to be consistently below the NAAQS. Table D.2-1 provides primary (public health, including health of "sensitive" populations) and secondary (public welfare, e.g., protection against vegetation and materials damage, decrease in visibility) NAAQS for each criteria air pollutant. However, if the pollution limits are exceeded for several consecutive years, the EPA will designate an area as being in nonattainment. The area will subsequently be subject to more stringent new or modified source regulatory requirements.

Areas can be in attainment for some pollutants, while being in nonattainment for others. Some areas are designated as "maintenance" areas. These are regions that were initially designated as nonattainment or unclassifiable and have since attained compliance with the NAAQS. Some designated nonattainment areas for some pollutants include classifications identifying the level of severity or degree of nonattainment. For example, the 1997 8-hr O_3 standard of 0.08 ppm has six separate classification levels, ranging from "marginal" to "extreme".[b] Further details on the nonattainment designations, including the classification levels, can be found in EPA's "Green Book" (http://www.epa.gov/oaqps001/greenbk/).

(b) Extreme—Area has a design value of 0.187 ppm and above. Severe 17—Area has a design value of 0.127 up to but not including 0.187 ppm. Severe 15—Area has a design value of 0.120 up to but not including 0.127 ppm. Serious—Area has a design value of 0.107 up to but not including 0.120 ppm. Moderate—Area has a design value of 0.092 up to but not including 0.107 ppm. Marginal—Area has a design value of 0.085 up to but not including 0.092 ppm.

Table D.2-1. National Ambient Air Quality Standards (NAAQS)

Pollutant[a]	Averaging Time	NAAQS[b] Value	Type[c]
SO_2	1-hour	75 ppb[d]	P
	3-hour	0.5 ppm	S
NO_2	1-hour	100 ppb	P
	Annual	0.053 ppm (53 ppb)	P, S
CO	1-hour	35 ppm	P
	8-hour	9 ppm	P
O_3	8-hour	0.075 ppm[e]	P, S
PM_{10}	24-hour	150 μg/m^3	P, S
$PM_{2.5}$	24-hour	35 μg/m^3	P, S
	Annual	15 μg/m^3	P, S
Pb	Rolling 3-month	0.15 μg/m^3[f]	P, S

(a) Notation: CO = carbon monoxide; NO_2 = nitrogen dioxide; O_3 = ozone; Pb =lead; $PM_{2.5}$ = particulate matter ≤ 2.5 μm; PM_{10} = particulate matter ≤ 10 μm; and SO_2 = sulfur dioxide.

(b) Refer to 40 CFR Part 50 for detailed information on attainment determination and reference method for monitoring.

(c) P = Primary standard whose limits were set to protect public health; S = Secondary standard whose limits were set to protect public welfare.

(d) Final rule signed June 2, 2010. The 1971 annual and 24-hour SO_2 standards were revoked in that same rulemaking. However, these standards remain in effect until one year after an area is designated for the 2010 standard, except in areas designated nonattainment for the 1971 standards, where the 1971 standards remain in effect until implementation plans to attain or maintain the 2010 standard are approved.

(e) Final rule signed March 12, 2008. The 1997 ozone standard (0.08 ppm, annual fourth-highest daily maximum 8-hour concentration, averaged over 3 years) and related implementation rules remain in place. In 1997, EPA revoked the 1-hour ozone standard (0.12 ppm, not to be exceeded more than once per year) in all areas, although some areas have continued obligations under that standard ("anti-backsliding"). The 1-hour ozone standard is attained when the expected number of days per calendar year with maximum hourly average concentrations above 0.12 ppm is less than or equal to 1.

(f) Final rule signed October 15, 2008. The 1978 lead standard (1.5 μg/m^3 as a quarterly average) remains in effect until one year after an area is designated for the 2008 standard, except that in areas designated nonattainment for the 1978, the 1978 standard remains in effect until implementation plans to attain or maintain the 2008 standard are approved.

Source: EPA 2011a

To be reclassified from nonattainment to an attainment maintenance area, the Clean Air Act outlines several conditions that must be met, one of which is the development and EPA approval of a maintenance plan. Other conditions that States must meet before an area may be redesignated by the EPA include: (1) the area has monitored attainment of the air quality standard; (2) the EPA has determined that the improvement in air quality is due to permanent and enforceable reductions in emissions; (3) the State has submitted and EPA has approved a maintenance plan for the area; and (4) the area has met all other applicable Clean Air Act requirements. The EPA may approve or deny the redesignation request based on air monitoring information, the activities listed in the State Implementation Plan, and the comments submitted by the public.

The maintenance plan must demonstrate continued compliance, considering projected growth, for a period of ten years. If outdoor air monitors record a violation of the standard, the maintenance plan must include a commitment to determine appropriate measures to address the cause of the violation. This plan must specify measures that will be used in the area to maintain compliance with the NAAQS. The plan must include controls the area will employ to ensure emissions remain below certain levels and contingency measures to ensure prompt correction of any NAAQS violations. The U.S. Nuclear Regulatory Commission (NRC) will ensure coordination of the licensee with the appropriate EPA Regional Office and/or State air quality office before any plants begin major construction or refurbishment activities.

D.3 Geologic Environment

D.3.1 Description of Affected Resources and Region of Influence

The geologic environment of nuclear power plant sites was not addressed in the 1996 GEIS, but geology and soils are included in this GEIS update. An understanding of such geologic and soil conditions, as well as the presence of geologic hazards, has been well established at all nuclear power plants during the current licensing term. Changes in the potential for hazards, such as earthquakes, are not within the scope of this GEIS because any such changes during the period of extended operation would not be the result of nuclear reactor operations. The geologic and soil resources considered in this GEIS are those that could be affected by an additional 20 years of reactor operation and by refurbishment activities within the nuclear power plant site property boundaries and nearby offsite areas. Because land and soil disturbance during license renewal could occur in undisturbed and undeveloped areas either onsite or possibly offsite, the locations of power plants relative to areas of important farmland soils (e.g., prime farmland) were considered. In addition, the region of potentially affected geologic resources considered extends to offsite areas because the presence of a nuclear power plant may restrict rock, mineral, and fossil fuel extraction operations beyond the site boundaries.

D.3.2 Description of Impact Assessment

Geologic and soil resources could be affected by construction or refurbishment projects during the license renewal term or subsequently during plant decommissioning. These actions would include activities that disturb surface soils, sediments, and underlying geologic strata, resulting in such effects as erosion, loss of soil resources, and increased suspended solids in nearby surface water bodies.

All published SEISs were reviewed for new and significant information pertaining to geologic and soil impacts from continued operations and refurbishment, but none was noted. The magnitude of the impact of potential ground-disturbing activities on geology and soils and local geologic resources would depend on site-specific factors such as the nature of geologic strata and soils, facility location, construction planning, and site-specific resource mapping.

D.4 Water Resources

D.4.1 Description of Affected Resources and Region of Influence

Most of the nuclear power plants are located near significant surface water bodies that are either natural or man-made. Therefore, the region of influence for water resources includes those on and adjacent to each nuclear power plant site that could be impacted by water withdrawals, effluent discharges, and spills or stormwater runoff associated with continued operations and refurbishment activities. Thus, the surface water resources considered includes those onsite, downstream of the site (in the case of river settings), or throughout some portion of a body of water (in the case of an ocean, lake or Great Lake, bay, reservoir, or pond) adjacent to the site. The region of influence for groundwater impacts includes areas both onsite (local water table) and offsite (regional aquifer).

D.4.2 Description of Impact Assessment

Sources of information about surface-water and groundwater issues regarding water use, water conflicts, and water quality included the 1996 GEIS, plant-specific supplements to the GEIS, and the decommissioning GEIS update (NRC 2002). All published SEISs were reviewed for new and significant information pertaining to water issues.

To analyze the condenser flow rate requirements and consumptive loss associated with specific categories of cooling system technologies (see Section 3.5.1.1 in this GEIS), data from the 1996 GEIS and a U.S. Geological Survey report were compiled. The flow rates and consumptive loss rates were normalized to a specific power capacity to allow comparisons.

Permitting requirements related to surface water withdrawal and groundwater use were summarized, and recent information was reviewed to assess water use conflicts and drought effects on rivers.

The evaluation of new and significant information related to water resources impacts led to several new water issues being considered in this GEIS revision. For example, the impacts of dredging were addressed by reviewing information on dredging operations and permitting requirements in SEISs prepared since issuance of the 1996 GEIS. The effects of general groundwater and soil contamination stemming from spills, leaks, and general industrial practices at power plants were evaluated through review of plant-specific SEISs and supporting documents. The impacts of radionuclide leaks, particularly tritium, were summarized based on a recent NRC summary report of tritium incidents (NRC 2006). A related document by the nuclear industry (NEI 2007) pertaining to assessment and monitoring was also reviewed.

D.5 Ecological Resources

D.5.1 Description of Affected Resources and Region of Influence

Terrestrial resources potentially affected by nuclear power plant operations during the license renewal term were determined at a broad level by obtaining the Level III ecoregion data (EPA 2007) (Table D.5-1) and land cover data (USGS 2003) for the vicinity of each power plant. An ecoregion describes a broad landscape in which the ecosystems have a general similarity. It can be characterized by the spatial pattern and composition of biotic and abiotic features, such as vegetation, wildlife, physiography, climate, soils, and hydrology (CEC 1997). The Level I ecoregions of the United States in which the nuclear power plants are located are shown in Figure D.5-1. Each ecoregion is subdivided into subregions. Level III ecoregions range from the warm, arid Sonoran Basin and Range ecoregion with cactus-shrub habitats, in which the Palo Verde plant in Arizona is located, to the cool, moist Northeastern Highlands ecoregion with northern hardwood and spruce-fir forests, which contains the Indian Point plant in New York. Level III ecoregions in the vicinity of the operating nuclear plants are presented in Table D.5-2. The region of influence for each power plant was considered to be the area within a radius of 5 mi as well as the transmission line ROWs associated with each power plant.

In the vicinity of the nuclear plants, an average of 25 percent of the land area is forested, 5 percent is grassland, and 4 percent is shrubland, as determined from land cover data. The land area around 10 plants is mostly forested (exceeding 50 percent of the land cover), and around 2 plants, it is mostly shrubland. (For no plants is it mostly grassland.) Agricultural lands are also present in the vicinity of many of the nuclear plants. An average of 23 percent of the area around all plants is used for crop production, and the area around 9 nuclear plants is mostly agricultural (greater than 30 percent land cover). Wetland types within 5 mi of each

Table D.5-1. Level I Ecoregions and Corresponding Level III Ecoregions That Occur in the Vicinity of U.S. Commercial Nuclear Power Plants

Level I Ecoregion	Level III Ecoregion	Level III Description
Eastern Temperate Forests	Arkansas Valley	Forest, pasture, cropland; bottomland deciduous forest on floodplains
	Atlantic Coastal Pine Barrens	Northeastern oak-pine woodland; marsh, swamp, floodplain forest along tidal rivers, freshwater marsh; agriculture; dunes, barrier islands
	Central Corn Belt Plains	Cropland; tallgrass prairie, oak-hickory forest
	Driftless Area	Agriculture; hardwood forest
	Eastern Great Lakes and Hudson Lowlands	Agriculture; northern hardwood forest
	Erie Drift Plain	Agriculture; maple-beech-birch forest; wetlands
	Huron/Erie Lake Plains	Cropland; oak savanna on dunes/beach ridges, elm-ash swamp, beech forest
	Interior Plateau	Oak-hickory forest, cropland, pasture; bluestem prairie, cedar glades
	Interior River Valleys and Hills	Cropland; pasture; forested valley slopes, bottomland deciduous forest, swamp forest, mixed oak forest, oak-hickory forest
	Middle Atlantic Coastal Plain	Pine forest, swamp, marsh, estuaries; oak, gum, cypress near rivers; cropland; dunes, barrier islands
	Mississippi Alluvial Plain	Cropland; bottomland deciduous forest
	Mississippi Valley Loess Plains	Cropland; oak-hickory forest and oak-hickory-pine forest
	North Central Hardwood Forests	Mosaic of coniferous forest and northern hardwood forest, wetlands and lakes, cropland, pasture
	Northeastern Coastal Zone	Northern hardwood forest, spruce-fir forest, lakes
	Northern Piedmont	Appalachian oak forest, cropland

Table D.5-1. (cont.)

Level I Ecoregion	Level III Ecoregion	Level III Description
Eastern Temperate Forests (cont.)	Piedmont	Oak-hickory-pine woodland; cropland
	Ridge and Valley	Appalachian oak forest, oak-hickory-pine forest, agriculture
	S. Michigan/N. Indiana Drift Plains	Lakes, marsh; agriculture; oak-hickory forest, northern swamp forest, beech forest
	Southeastern Plains	Mosaic of cropland, pasture, woodland, mixed forest
	Southeastern Wisconsin Till Plains	Agriculture; mosaic of hardwood forest, oak savanna, tallgrass prairie
	Southern Coastal Plain	Coastal lagoons, marsh, swamp, barrier islands; pine, oak-gum-cypress forest; citrus groves, pasture; lakes
	Western Allegheny Plateau	Mixed mesophytic forest, mixed oak forest; pasture, cropland
Great Plains	Central Irregular Plains	Mosaic of grassland, wide riparian forest; cropland
	Cross Timbers	Rangeland, pasture; little bluestem grassland with scattered oaks
	Western Corn Belt Plains	Cropland, pasture; tallgrass prairie; narrow riparian forest
	Western Gulf Coastal Plain	Grassland, cropland
North American Deserts	Columbia Plateau	Arid sagebrush steppe and grassland; agriculture
	Sonoran Basin and Range	Hot climate; creosote bush-bur sage; large areas of palo verde-cactus shrub and giant saguaro cactus
Mediterranean California	Southern and Central California Chaparral and Oak Woodlands	Mediterranean climate: hot dry summers, cool moist winters
Northern Forests	Northeastern Highlands	Northern hardwood forest, spruce-fir forest; lakes
Tropical Wet Forests	Southern Florida Coastal Plain	Frost-free climate; flat plains with wet soils; marshland, swamp, everglades, palmetto prairie

Source: EPA 2007

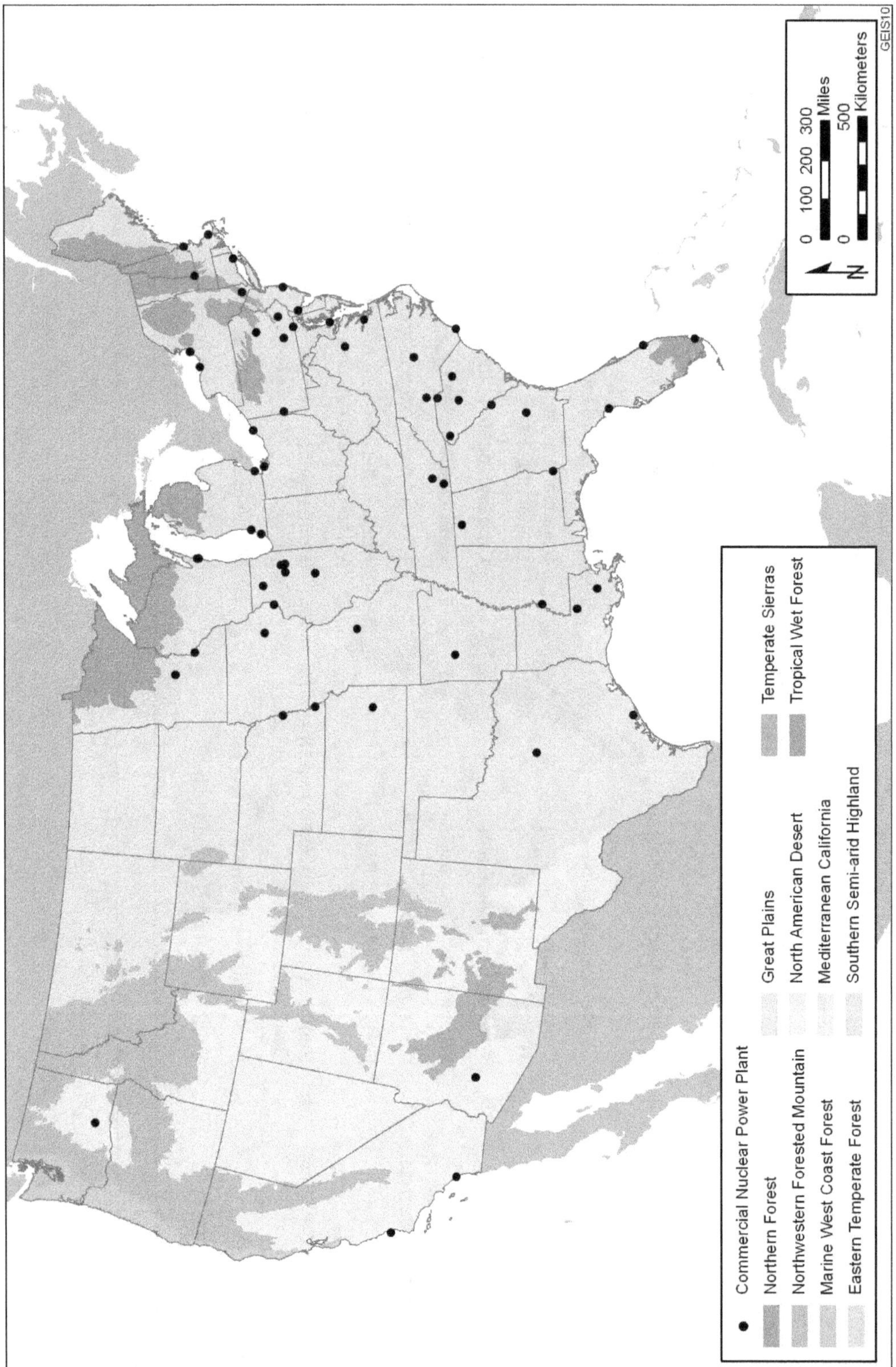

Figure D.5-1. Level I Ecoregions of the United States (CEC 2006)

Table D.5-2. Ecoregions in the Vicinity of Operating Nuclear Power Plants

Site Name	Level I Description	Level III Ecoregion(s)
Arkansas	Eastern Temperate Forests	Arkansas Valley
Beaver Valley	Eastern Temperate Forests	Western Allegheny Plateau
Braidwood	Eastern Temperate Forests	Central Corn Belt Plains
Browns Ferry	Eastern Temperate Forests	Interior Plateau
Brunswick	Eastern Temperate Forests	Middle Atlantic Coastal Plain
Byron	Eastern Temperate Forests	Central Corn Belt Plains
Callaway	Eastern Temperate Forests	Interior River Valleys and Hills
Calvert Cliffs	Eastern Temperate Forests	Southeastern Plains, Middle Atlantic Coastal Plain
Catawba	Eastern Temperate Forests	Piedmont
Clinton	Eastern Temperate Forests	Central Corn Belt Plains
Columbia	North American Deserts	Columbia Plateau
Comanche Peak	Great Plains	Cross Timbers
Cooper	Great Plains	Western Corn Belt Plains
Crystal River	Eastern Temperate Forests	Southern Coastal Plain
D.C. Cook	Eastern Temperate Forests	S. Michigan/N. Indiana Drift Plains
Davis Besse	Eastern Temperate Forests	Huron/Erie Lake Plains
Diablo Canyon	Mediterranean California	Southern and Central California Chaparral and Oak Woodlands
Dresden	Eastern Temperate Forests	Central Corn Belt Plains
Duane Arnold	Great Plains	Western Corn Belt Plains
Farley	Eastern Temperate Forests	Southeastern Plains
Fermi	Eastern Temperate Forests	Huron/Erie Lake Plains
FitzPatrick	Eastern Temperate Forests	Eastern Great Lakes and Hudson Lowlands

Table D.5-2. (cont.)

Site Name	Level I Description	Level III Ecoregion(s)
Fort Calhoun	Great Plains	Western Corn Belt Plains
Ginna	Eastern Temperate Forests	Eastern Great Lakes and Hudson Lowlands
Grand Gulf	Eastern Temperate Forests	Mississippi Valley Loess Plains, Mississippi Alluvial Plain
Harris	Eastern Temperate Forests	Piedmont, Southeastern Plains
Hatch	Eastern Temperate Forests	Southeastern Plains, Southern Coastal Plain
Hope Creek	Eastern Temperate Forests	Middle Atlantic Coastal Plain
Indian Point	Northern Forests	Northeastern Highlands
Kewaunee	Eastern Temperate Forests	Southeastern Wisconsin Till Plains
LaSalle	Eastern Temperate Forests	Central Corn Belt Plains
Limerick	Eastern Temperate Forests	Northern Piedmont
McGuire	Eastern Temperate Forests	Piedmont
Millstone	Eastern Temperate Forests	Northeastern Coastal Zone
Monticello	Eastern Temperate Forests	North Central Hardwood Forests
Nine Mile Point	Eastern Temperate Forests	Eastern Great Lakes and Hudson Lowlands
North Anna	Eastern Temperate Forests	Piedmont
Oconee	Eastern Temperate Forests	Piedmont
Oyster Creek	Eastern Temperate Forests	Atlantic Coastal Pine Barrens
Palisades	Eastern Temperate Forests	S. Michigan/N. Indiana Drift Plains
Palo Verde	North American Deserts	Sonoran Basin and Range
Peach Bottom	Eastern Temperate Forests	Northern Piedmont
Perry	Eastern Temperate Forests	Eastern Great Lakes and Hudson Lowlands, Erie Drift Plain
Pilgrim	Eastern Temperate Forests	Atlantic Coastal Pine Barrens, Northeastern Coastal Zone
Point Beach	Eastern Temperate Forests	Southeastern Wisconsin Till Plains

Table D.5-2. (cont.)

Site Name	Level I Description	Level III Ecoregion(s)
Prairie Island	Eastern Temperate Forests	Driftless Area
Quad Cities	Eastern Temperate Forests and Great Plains	Interior River Valleys and Hills, Western Corn Belt Plains, Central Corn Belt Plains
River Bend	Eastern Temperate Forests	Mississippi Valley Loess Plains, Mississippi Alluvial Plain
Robinson	Eastern Temperate Forests	Southeastern Plains
Salem	Eastern Temperate Forests	Middle Atlantic Coastal Plain
San Onofre	Mediterranean California	Southern and Central California Chaparral and Oak Woodlands
Seabrook	Eastern Temperate Forests	Northeastern Coastal Zone
Sequoyah	Eastern Temperate Forests	Ridge and Valley
South Texas	Great Plains	Western Gulf Coastal Plain
St. Lucie	Eastern Temperate Forests	Southern Coastal Plain
Summer	Eastern Temperate Forests	Piedmont
Surry	Eastern Temperate Forests	Middle Atlantic Coastal Plain, Southeastern Plains
Susquehanna	Eastern Temperate Forests	Ridge and Valley
Three Mile Island	Eastern Temperate Forests	Northern Piedmont
Turkey Point	Tropical Wet Forests	Southern Florida Coastal Plain
Vermont Yankee	Eastern Temperate Forests and Northern Forests	Northeastern Coastal Zone, Northeastern Highlands
Vogtle	Eastern Temperate Forests	Southeastern Plains
Waterford	Eastern Temperate Forests	Mississippi Alluvial Plain
Watts Bar	Eastern Temperate Forests	Ridge and Valley
Wolf Creek	Great Plains	Central Irregular Plains

Source: EPA 2007

power plant were determined by obtaining National Wetland Inventory (NWI) data (USFWS 2007) (Table D.5-3). When NWI data were not available, the land cover data were used. Open water areas were assigned to NWI classification on the basis of NWI classification methodology.

Aquatic habitats and the types of aquatic organisms (including special status species) that could be affected by nuclear power plant operations during the license renewal term were determined at a broad level on the basis of the location of the plant and the source of cooling water used by the plant. In cases where cooling systems could affect more than one type of system (e.g., freshwater and estuarine), impacts to both systems were considered in the analysis. Similarly, the potential for migratory aquatic species to be affected by a particular plant was based on reported occurrences of such species in waters used to supply cooling water. Plants that use estuarine or marine water sources for cooling or plants that use freshwater cooling water sources with a potential for containing migratory life stages of Federally managed fishery species were assumed to have a potential for affecting essential fish habitat. In general, existing impingement, entrainment, and thermal impacts on aquatic organisms from cooling water systems were considered to be lower for plants with cooling towers when operating in a closed-cycle mode because those plants withdraw smaller volumes of water for cooling and have comparatively smaller thermal plumes.

Additional information regarding terrestrial and aquatic resources in the vicinity of specific nuclear power plants was obtained from scientific articles and reports, from recently completed SEISs, and from environmental reports (ERs) included as part of the applications submitted by applicants for renewal of reactor licenses. Information from these sources was used to describe the general types of nuclear plant interactions with ecological resources and illustrate impact types observed at the nuclear plants. In some cases, information provided in the 1996 GEIS (NRC 1996) was used to describe the affected environment.

D.5.2 Description of Impact Assessment

A wide range of issues (Table 2.4-1) related to potential impacts of license renewal on ecological resources were evaluated by considering how continuation of operations would affect ecological resources compared to the current condition. Although the ecological impacts associated with plant decommissioning have been previously evaluated by the NRC (2002), the ecological impacts associated with delaying decommissioning because of license renewal were considered as part of the proposed action (Section 4.1.3.5). Potential impacts to terrestrial and aquatic resources were evaluated, in part, by a review of published literature related to the impacting factors associated with operations and associated construction and refurbishment actions during the license renewal term. Although some of the impacts identified were specific to nuclear power plant operation (e.g., effects of radionuclides on biota), impacts associated with non-nuclear power plants also were reviewed (e.g., the effects of bird collisions

Table D.5-3. Percent of Area Occupied by Wetland and Deepwater Habitats Within 5 Miles of Operating Nuclear Power Plants

Site	Percent of Area Occupied by Wetland and Deepwater Habitat Types								
	Estuarine and Marine Deepwater[b]	Estuarine and Marine Wetland	Lacustrine	Palustrine Pond[c]	Riverine	Palustrine Emergent Wetland	Palustrine Forested/Shrub Wetland	Other	Total Wetland[d]
Arkansas[a]	0.0	0.0	11.4	0.0	0.0	0.1	0.2	0.0	11.7
Beaver Valley	0.0	0.0	1.6	0.2	3.5	0.1	0.1	0.0	5.5
Braidwood	0.0	0.0	7.6	1.6	0.4	1.0	0.8	0.0	11.4
Browns Ferry	0.0	0.0	26.6	0.2	0.0	0.7	14.7	0.0	42.2
Brunswick	23.6	15.3	0.6	0.4	0.2	0.5	19.9	0.1	36.9
Byron	0.0	0.0	1.9	0.1	0.0	0.6	0.9	0.0	3.6
Callaway	0.0	0.0	0.3	0.5	1.1	0.8	1.7	0.0	4.5
Calvert Cliffs	52.8	0.5	0.0	0.1	0.0	0.3	2.3	0.0	3.3
Catawba	0.0	0.0	12.0	0.3	0.2	0.0	0.3	0.0	12.9
Clinton	0.0	0.0	8.3	0.1	0.0	0.1	0.4	0.0	9.0
Columbia	0.0	0.0	5.4	0.0	0.0	0.1	0.1	0.0	5.6
Comanche Peak[a]	0.0	0.0	6.6	0.2	0.4	0.0	1.6	0.0	8.8
Cooper	0.0	0.0	0.1	0.3	2.7	0.7	3.1	0.0	6.8
Crystal River	39.1	13.9	0.1	0.3	0.2	2.1	9.6	0.0	26.1
Cook	0.0	0.0	50.6	0.3	0.0	0.5	2.2	0.0	53.6
Davis Besse	0.0	0.0	53.7	0.8	2.7	7.4	2.0	0.0	66.6
Diablo Canyon	53.8	0.3	0.0	0.0	0.1	0.0	0.5	0.0	0.9
Dresden	0.0	0.0	10.8	1.5	1.2	5.1	3.4	0.0	22.0
Duane Arnold	0.0	0.0	1.0	0.5	1.6	1.2	7.5	0.0	11.7
Farley	0.0	0.0	1.5	0.5	0.0	1.0	10.1	0.0	13.1
Fermi	0.0	0.0	53.3	0.4	0.8	2.0	1.4	0.0	57.9
FitzPatrick[a]	0.0	0.0	60.5	0.0	0.0	0.2	4.7	0.0	65.4
Fort Calhoun	0.0	0.0	1.8	0.3	1.7	0.9	1.6	0.0	6.3

Table D.5-3. (cont.)

Site	Percent of Area Occupied by Wetland and Deepwater Habitat Types								
	Estuarine and Marine Deepwater[b]	Estuarine and Marine Wetland	Lacustrine	Palustrine Pond[c]	Riverine	Palustrine Emergent Wetland	Palustrine Forested/Shrub Wetland	Other	Total Wetland[d]
Ginna[a]	0.0	0.0	61.7	0.1	0.0	0.1	1.2	0.0	63.1
Grand Gulf[a]	0.0	0.0	3.1	0.0	8.8	0.7	26.8	0.0	39.4
Harris	0.0	0.0	9.3	0.3	0.0	0.0	3.5	0.0	13.2
Hatch	0.0	0.0	0.0	0.8	2.3	0.5	20.3	0.0	23.9
Hope Creek	44.1	34.6	0.0	0.4	0.0	1.7	1.4	0.1	38.3
Indian Point	14.7	0.6	0.6	0.8	0.2	0.4	1.8	0.0	4.3
Kewaunee[a]	0.0	0.0	46.6	0.0	0.0	1.0	4.3	0.0	51.9
LaSalle	0.0	0.0	4.5	0.3	0.0	0.1	0.1	0.0	4.9
Limerick	0.0	0.0	0.0	0.3	1.2	0.1	0.4	0.0	2.0
McGuire	0.0	0.0	19.5	0.2	0.0	0.1	1.5	0.0	21.4
Millstone	49.1	1.7	0.3	0.2	0.0	0.1	2.1	0.0	4.3
Monticello	0.0	0.0	2.8	0.7	1.4	4.3	2.5	0.0	11.8
Nine Mile Point[a]	0.0	0.0	60.8	0.0	0.0	0.2	4.6	0.0	65.7
North Anna	0.0	0.0	18.3	0.2	0.1	0.1	2.4	0.0	21.1
Oconee[a]	0.0	0.0	21.8	0.2	0.0	0.0	0.4	0.0	22.3
Oyster Creek	27.0	3.9	0.2	0.2	0.0	0.6	13.0	0.0	18.0
Palisades	0.0	0.0	48.9	0.2	0.0	1.1	7.8	0.1	58.1
Palo Verde[a]	0.0	0.0	1.1	0.0	0.0	0.0	0.1	0.0	1.2
Peach Bottom	0.0	0.0	13.8	0.2	0.3	0.1	0.1	0.0	14.5
Perry[a]	0.0	0.0	48.6	0.1	0.1	0.0	0.5	0.0	49.3
Pilgrim	60.8	1.0	0.8	0.3	0.0	0.1	1.4	0.0	3.6
Point Beach[a]	0.0	0.0	45.3	0.0	0.0	1.2	7.5	0.0	54.0
Prairie Island[a]	0.0	0.0	13.2	1.0	0.1	4.4	13.2	0.0	31.9
Quad Cities	0.0	0.0	8.9	0.7	0.2	2.0	10.3	0.0	22.2

Table D.5-3. (cont.)

Site	Percent of Area Occupied by Wetland and Deepwater Habitat Types								
	Estuarine and Marine Deepwater[b]	Estuarine and Marine Wetland	Lacustrine	Palustrine Pond[c]	Riverine	Palustrine Emergent Wetland	Palustrine Forested/Shrub Wetland	Other	Total Wetland[d]
River Bend[a]	0.0	0.0	0.7	0.7	6.2	1.3	32.7	0.0	41.6
Robinson	0.0	0.0	4.4	0.4	0.0	0.3	8.3	0.0	13.5
Salem	45.0	35.6	0.0	0.4	0.0	1.5	1.3	0.1	39.0
San Onofre	48.1	0.5	0.0	0.0	0.7	0.1	1.3	0.0	2.6
Seabrook	24.3	12.8	0.1	0.3	0.0	0.7	6.9	0.0	20.9
Sequoyah	0.0	0.0	14.7	0.4	0.0	0.0	0.1	0.0	15.1
South Texas	0.0	0.0	14.2	0.2	0.8	5.0	3.1	0.0	23.3
St. Lucie	65.8	4.3	0.0	0.5	0.1	7.0	0.1	0.0	11.9
Summer	0.0	0.0	17.1	0.2	0.6	0.3	2.0	0.0	20.2
Surry	33.8	2.6	1.4	0.2	17.3	3.2	2.4	0.0	27.0
Susquehanna	0.0	0.0	0.2	0.3	3.1	0.1	1.0	0.0	4.6
Three Mile Island	0.0	0.0	0.0	0.3	10.6	0.1	0.3	0.0	11.4
Turkey Point	53.7	12.9	0.0	0.0	0.1	15.7	9.1	0.0	37.8
Vermont Yankee	0.0	0.0	2.7	0.3	1.1	0.6	1.4	0.0	6.1
Vogtle	0.0	0.0	0.3	0.3	1.0	1.6	24.3	0.0	27.4
Waterford	0.0	0.0	1.6	0.9	7.6	11.8	45.8	0.2	67.8
Watts Bar	0.0	0.0	9.2	0.2	0.3	0.2	1.1	0.0	10.9
Wolf Creek[a]	0.0	0.0	13.0	0.7	0.4	0.2	0.6	0.0	14.8

(a) Data were derived, at least in part, from the National Land Cover Database (USGS 2003) and were assigned to NWI categories.
(b) Deepwater habitats are permanently flooded and lie below the deepwater/wetland boundary (Cowardin et al 1979).
(c) Includes the Aquatic Bed and Unconsolidated Bottom wetland classes.
(d) Does not include deepwater habitats.
Source: National Wetlands Inventory (USFWS 2007), except where noted.

with natural draft cooling towers or electric transmission lines, or the effects of impingement, entrainment, and thermal stress on fish and other aquatic organisms). In addition, recently completed SEISs were reviewed for impact evaluations and the presentation of new and potentially significant information on the impacts of plant operations during the renewal term.

The potential impacts of radionuclide exposure on terrestrial and aquatic biota at nuclear power plants were evaluated by reviewing Radiological Environmental Monitoring Program reports (primarily annual radiological environmental operating reports) for 15 power plants selected to represent a range of radionuclide concentrations in the environmental media, including plants identified as having high annual worker total effective dose equivalent values or public exposures. In instances where a site's sediment or soil concentration for a particular nuclide was below the lower limit of detection (LLD), the LLD was substituted as the concentration for that media type, thereby resulting in a conservative estimate (i.e., more likely to identify a potential for negative impacts to biota) of potential exposure. The radionuclide concentrations in water, sediment, and soil were then input to the RESRAD-BIOTA dose evaluation model (DOE 2004) to estimate the dose rates for terrestrial and aquatic biota.

The RESRAD-BIOTA code was developed at Argonne National Laboratory based on the U.S. Department of Energy's (DOE's) graded approach to biota dose evaluation (DOE 2002). There are three levels provided by the RESRAD-BIOTA code corresponding to the graded approach to biota dose evaluation. The evaluation presented in Section 4.1.1.5.1 was conducted using RESRAD-BIOTA Level 2, which was necessary for dose modeling. Because the LLDs for water samples were relatively high compared to the Biota Concentration Guide, water radionuclide concentrations below the LLD were estimated using the partition coefficient (K_d value) provided in the RESRAD-BIOTA code.

For all ecological receptors, default bioaccumulation factors and dose limits were used. Radionuclides at each site were evaluated by comparing the sum of the total estimated dose to the default dose limits (riparian animal, 0.1 rad/d; terrestrial animal, 0.1 rad/d; terrestrial plant, 1.0 rad/d; aquatic organisms, 1.0 rad/d). Estimated doses that were less than the dose limit were determined to represent an acceptable radiological risk to the receptor, whereas estimated doses above the dose limit were determined to represent an unacceptable radiological risk to the receptor. More information about the RESRAD-BIOTA code, including instructions for using the model, can be found at http://web.ead.anl.gov/resrad/documents/.

The potential impacts of continued operation of the cooling systems on terrestrial biota at nuclear power plants were evaluated by reviewing published site-specific reports to gather information on the types and concentrations of contaminants released from the cooling systems into the environment and comparing those concentrations to regulatory guidelines to determine whether the contaminants associated with cooling system operation posed any risk to terrestrial resources. Specifically, radiological effluent release reports (RERRs), ERs, and recently prepared SEISs for eight nuclear power plants were reviewed to identify the types and concentrations of contaminants associated with the operation of the cooling systems. The eight nuclear power plants were selected to represent different cooling systems and contaminants. Water concentrations were reported in the RERRs, ERs, or SEISs for only two contaminants: chlorine and tritium. The maximum reported concentrations for both contaminants from the site-specific reports were compared to regulatory guidelines and the results from laboratory experiments. Maximum site-specific concentrations below the lowest observed effects level (LOEL) or below the recommended regulatory guideline were considered to represent an acceptable risk to terrestrial resources. Maximum site-specific concentrations above the LOEL or above the recommended regulatory guideline were considered to represent an unacceptable risk to terrestrial resources. Potential effects of contaminants introduced into aquatic environments from cooling water systems were evaluated by reviewing the SEISs that have been previously completed and scientific literature pertaining to potential and observed effects of contaminants and biocides used for maintenance of cooling water systems.

D.6 Historic and Cultural Resources

D.6.1 Description of Affected Resources and Region of Influence

In this revision, the term “historic properties” is used when discussing Section 106 compliance activities and “historic and cultural resources” is used when generically referencing the resource. The NRC considers historic and cultural resources as an all-inclusive term that includes prehistoric, historic, and traditional cultural properties. In addition, while NHPA requires agencies to take into account the effects of their undertakings on historic properties, NEPA requires the consideration of the cultural environment; thus the issue is termed “Historic and Cultural Resources” (see NEPA Statute Sec. 101 42 USC 4331(b) 4). The NRC coordinates Section 106 requirements through the NEPA process.

In determining affected historic and cultural resources, the region of influence is the Area of Potential Effect (APE). The license renewal APE is the area that may be impacted by land disturbing, or other operational, activities associated with continued plant operations and maintenance during the license renewal term and/or refurbishment. The APE typically encompasses the nuclear power plant site, its immediate environs including viewshed, and the transmission lines within this scope of review (see Sections 3.1.1 and 3.1.6.5 in this GEIS). The APE may extend beyond the nuclear plant site and transmission lines when these activities may affect historic and cultural resources. This determination is made irrespective of land ownership or control. The NRC must identify historic and cultural resources occurring within the defined APE.

Historic and cultural resources can include physical remains of past human activity that have historic or cultural meaning. They include archaeological sites (e.g., prehistoric campsites and villages), historic era resources (e.g., farmsteads, forts, and canals), and traditional cultural properties (e.g., resource collection areas and sacred areas). Historic and cultural resources that are eligible for listing on the NRHP are considered historic properties that are evaluated under Section 106 of the NHPA. Historic properties that could be affected by license renewal and that are included in the region of influence include both those found in areas within the plant property and areas outside the property that would be affected by plant activities. In most cases, license renewal activities will be confined to the current property boundaries and the transmission line ROW up to the first substation. Continued operations, refurbishment, and decommissioning activities may affect currently undeveloped portions of plant property. While some portions of the nuclear power plant site and transmission line ROW were heavily disturbed during the power plant construction, it is expected that some historic and cultural resources would remain in less disturbed portions of the site.

D.6.2 Description of Impact Assessment

Cultural resources were identified as resources to be considered for license renewal in the 1996 GEIS (NRC 1996), where they were identified as a Category 2 issue (NRC 1996). The current assessment is in agreement with this categorization. Due to geographic, cultural, and historic differences, a site-specific assessment of historic and cultural resources must be performed. A sample review of 27 nuclear power plants revealed very few archaeological sites. Extensive ground-disturbing activities occurred during nuclear power plant construction, and much of the land immediately surrounding the power block was disturbed down to bedrock. This activity would have eliminated any potential for historic or cultural resources to be present in this portion of the power plant site. However, to effectively determine areas that could potentially contain historic and cultural resources, a survey of any previously disturbed areas of the nuclear power plant site must be conducted by qualified professionals and in consultation with the SHPO and other consulting parties. Plant activities that could affect historic and cultural resources during the renewal period include minor construction projects, maintenance actions, security improvements, landscaping activities, agricultural, and recreational activities. Impacts to historic and cultural resources from these activities are assessed in plant-specific supplements to the GEIS.

Potential impacts to historic and cultural resources were considered for decommissioning and were found to be SMALL (NRC 2002). The current assessment evaluates whether continued operations during the license renewal term would affect the conclusions in the decommissioning GEIS for both the plant and transmission lines. The current assessment is based on potential effects of continued operation and refurbishment activities on historic and cultural resources during the license renewal term.

D.7 Socioeconomics and Environmental Justice

D.7.1 Description of Affected Resources and Region of Influence

The impacts of nuclear power plant operations and refurbishment occur at the local level, in the county in which a plant is located, at the regional level, in the counties in which the majority of permanent plant employees reside, and at the State level. The definition of the region around each nuclear plant is based on employee residential location data and the location of vendors providing materials, equipment, and services necessary for operation, maintenance, and any construction that might be required for refurbishment activities. The majority of the economic and tax revenue data used in the GEIS update was derived from a series of reports developed by the Nuclear Energy Institute (NEI 2003, 2004a,b,c,d, 2005a,b, 2006a,b,c), which was used to describe the socioeconomic environment in the region in which a sample of 11 nuclear plants are located, and for the estimation of impacts of each plant at the local and State levels (Table D.7-1).

Table D.7-1. Definition of Local Areas and Regions at 11 Nuclear Plants

Plant	Counties in Local Area	Additional Counties in Region	State
Diablo Canyon	San Luis Obispo	None	California
Grand Gulf	Warren and Claiborne	Hinds, Franklin, Copiah, Adams	Mississippi
Indian Point	Westchester, Duchess, Orange, Putnam and Rockland	None	New York
Limerick	Montgomery	Berks, Chester	Pennsylvania
Millstone	New London	None	Connecticut
Oconee	Anderson, Greenville, Oconee and Pickens	None	South Carolina
Palo Verde	Maricopa	None	California
Peach Bottom	York	Lancaster, Chester	Pennsylvania
Susquehanna	Luzerne	None	Pennsylvania
Three Mile Island	Dauphin	Lancaster, Lebanon, York, Cumberland, Perry	Pennsylvania
Wolf Creek	Coffey	Lyon, Franklin, Anderson, Shawnee	Kansas

Sources: NEI 2003, 2004a,b,c,d, 2005a,b, 2006a,b,c

D.7.2 Estimation of Direct and Indirect Economic Effects

Nuclear power plant operations generate significant employment and expenditures at each plant site. Wage and salary and nonlabor expenditures create demand for a range of durable and nondurable goods provided by wholesalers and retailers, and also create demand for health and professional services and housing. Power plants also provide tax revenues for local and State governmental entities. In addition to employment, wages and salaries, and nonlabor expenditures directly associated with plant operations, power plants also produce indirect employment and income in the local and State economies as direct expenditures associated with wages and salaries, procurement, and tax revenues, which circulate through the economies, producing additional economic activity. The magnitude of the indirect economic impact of labor spending at each plant is determined by the extent to which plant employees live in the local area and region around each plant. The indirect impact of nonlabor expenditures is determined by the extent to which vendors of materials, equipment, and supplies are located in the local area and region.

Estimation of the indirect impact of nuclear power plants on local and State employment and income in NEI data was based on the use of regional economic multipliers in association with plant expenditure data for the construction and operations phases. Multipliers capture the indirect (offsite) effects of onsite activities associated with construction and operation of an activity or event. Expenditure data associated with the construction and operation of nuclear power plants were derived from individual utility sources, which provided the relevant construction and operating cost data for wages and salaries and nonlabor expenditures (procurement of materials, equipment, and services) and tax revenues.

Expenditure data in the NEI reports were mapped into the relevant North American Industry Classification System codes for use with multipliers from an IMPLAN model specified for the local area and State in which each power plant is located. IMPLAN input-output economic models are based on economic accounts showing the flow of commodities to industries from producers and institutional consumers. The accounts also show consumption activities by workers and owners of capital and imports from outside the region. The IMPLAN model contains 528 sectors representing industries in agriculture, mining, construction, manufacturing, wholesale and retail trade, utilities, finance, insurance and real estate, and consumer and business services. The model also includes information for each sector on employee compensation; proprietary and property income; personal consumption expenditures; Federal, State, and local expenditures; inventory and capital formation; and imports and exports. More information on the IMPLAN model and data can be found in each NEI report (NEI 2003, 2004a,b,c,d, 2005a,b, 2006a,b,c).

In addition to NEI data on direct power plant employment, wage and salary spending, materials and equipment expenditures, and local and State tax revenues, NEI estimates of indirect employment and income impacts at the local and State levels associated with power plant labor and nonlabor expenditures and tax revenue spending were reported in the analysis of impacts in the GEIS update.

Impacts of plant operations and refurbishment are likely to vary according to the scale of employment and expenditures at each power plant and the type of economy in which each plant is located. To assess the impact of power plant size and location in the GEIS update, 11 power plants for which direct and indirect impacts were estimated by NEI were classified according to whether the economic structure in the locality and region around each plant is rural or semi-urban. Rural areas often have relatively simple economies, and agriculture is often the primary economic activity. Many of the industries that provide equipment and services important to power plant operations are largely absent in rural areas, which have smaller, less diversified labor markets, with often lower skilled, lower paying occupations. In addition to agriculture and related activities, in some locations there may also be a range of other activities, including resource extraction, manufacturing, and transportation industries that provide employment and income. In semi-urban areas, where economic structures are more complex than in rural areas, there are a wider range of industries and larger and more diverse labor markets. Semi-urban areas may also serve specialized economic functions, including maritime shipping, fishing, boatbuilding, recreation, and tourism and numerous locations featuring residential areas hosting second homes and retirement communities.

D.7.3 Environmental Justice Assessment Methods

Executive Order 12898, "Federal Actions to Address Environmental Justice in Minority Populations and Low-Income Populations," requests independent Federal agencies to incorporate environmental justice as part of their missions. Specifically, it requests them to address, as appropriate, any disproportionately high and adverse human health or environmental effects of their actions, programs, or policies on minority and low-income populations. The NRC voluntarily complies with this Executive Order. Additional guidance for undertaking environmental justice reviews is described in Section 3.10.

The analysis of the impacts of nuclear power plant operations and refurbishment during the license renewal term on environmental justice has three parts: (1) a description of the geographic distribution of low-income and minority populations in the affected area; (2) an assessment of whether the impacts of license renewal would produce impacts that are high and adverse; and (3) if impacts are high and adverse, a determination as to whether these impacts disproportionately affect minority and low-income populations.

The analysis considers minority and low-income populations who reside within a 50-mi (80-km) radius of a nuclear plant. Data on low-income and minority individuals are collected and analyzed at the census tract or census block group level.

Minority individuals are those who identify themselves as members of the following population groups: Hispanic or Latino, American Indian or Alaska Native, Asian, Black or African-American, Native Hawaiian or Other Pacific Islander, or two or more races. Beginning with the 2000 census, where appropriate, the census form allows individuals to designate multiple population group categories to reflect their ethnic or racial origin. In addition, persons who classify themselves as being of multiple racial origins may choose up to six racial groups as the basis of their racial origins. The term minority includes all persons, including those classifying themselves in multiple racial categories, except those who classify themselves as not of Hispanic origin and as White or Other Race.

Minority populations are identified when (1) the minority population of an affected area exceeds 50 percent or (2) the minority population percentage of the affected area is "meaningfully greater than" the minority population percentage in the general population or other appropriate unit of geographic analysis. Minority populations may be communities of individuals living in close geographic proximity to one another, or a geographically dispersed or transient set of individuals (e.g., migrant workers or American Indians) where the group experiences common conditions of environmental exposure or effect. The appropriate unit of geographic analysis may be a political jurisdiction, county, region, or State or other similar unit that is chosen so as not to artificially dilute or inflate the affected minority population.

Low-income individuals are those whose annual income falls below the poverty line. The poverty line takes into account family size and the age of individuals in the family. In 1999, for example, the poverty line for a family of five with three children below the age of 18 was $19,882. For any given family below the poverty line, all family members are considered as being below the poverty line for the purposes of analysis. Low-income populations in an affected area are identified with the annual statistical poverty thresholds from the Census Bureau's Current Population Reports, Series PB60. Low-income populations may be communities of individuals living in close geographic proximity to one another, or a set of individuals (e.g., migrant workers) where the group experiences common conditions of environmental exposure or effect.

Nuclear power plant license renewal could affect environmental justice if any adverse health and environmental impacts are significantly high, and if these impacts would disproportionately affect minority and low-income populations. If the analysis determines that health and environmental impacts are not significant, there can be no disproportionate impacts to minority and low-income populations. Disproportionately high and adverse human health effects occur when the risk or rate of exposure to an environmental hazard for a minority or low-income population is significant (as defined by the Council on Environmental Quality, CEQ) and appreciably exceeds the risk or exposure rate for the general population or for another appropriate comparison group. Disproportionately high environmental impacts that are significant (as defined by CEQ) are impacts or risks of impacts on the natural or physical environment in a low-income or minority community that appreciably exceed the environmental impact on the larger community. Such effects may include ecological, cultural, economic, or social impacts. Adverse environmental impacts are impacts that are determined to be both harmful and significant (as defined by CEQ). In assessing cultural and aesthetic environmental impacts, impacts that uniquely affect geographically dislocated or dispersed minority or low-income populations or American Indian Tribes are considered (CEQ 1997).

D.8 Human Health

D.8.1 Radiological Effects

Nuclear power plants produce electricity through a heat-generating process known as "fission," in which neutrons split uranium atoms to produce large amounts of energy. Any material that is capable of undergoing fission by neutrons in a self-sustaining chain reaction is called fissile material. The most common fissile isotopes are uranium-235 and plutonium-239. Neutrons whose energy distribution is in thermal equilibrium with the ambient medium are called thermal neutrons. When a thermal neutron strikes uranium-235, it splits the uranium atom into two isotopes with a smaller atomic weight (called fission products) and several neutrons (the mean number of neutrons per fission of uranium-235 is 2.5) and gamma rays. All fission products are radioactive and decay to form other radioactive isotopes. The amount of energy generated in a fission reaction is about 200 MeV, and this energy is distributed among fission products, neutrons, and fission gamma rays. Most of the energy generated in the nuclear fission process is dissipated as thermal energy and is converted into electrical energy in a nuclear power plant. Nuclear fission differs from other forms of radioactive decay in that it can be harnessed and controlled via a chain reaction in which neutrons released by each fission event trigger yet more events, which, in turn, release more neutrons and cause more fission.

In a nuclear reactor, a controlled sustained chain reaction is produced. The core of a nuclear reactor consists of fuel (containing uranium enriched in uranium-235), a moderator to slow down the neutrons released in fission, a coolant to remove the thermal energy, and control rods for controlling the chain reaction. In enriched uranium, the percent composition of uranium-235 is increased (2 to 5 percent) from its natural composition (about 0.7 percent) in uranium. The nuclear power plants in the United States use water as both a moderator and a coolant. During the fission process, a large inventory of radioactive fission products builds up within the fuel. Virtually all of the fission products are contained within the fuel pellets. The fuel pellets are enclosed in hollow metal rods (cladding), which are hermetically sealed to further prevent the release of fission products. However, a small fraction of the fission products migrate from the fuel rods and contaminate the reactor coolant. The primary system coolant also has radioactive contaminants as a result of neutron activation (a process by which a stable atom becomes radioactive after undergoing a reaction with a neutron). Neutrons also interact with structural materials inside the pressure vessel and with the pressure vessel itself and make those materials radioactive.

D.8.1.1 Background Information on Radiation

Atoms are the basic building blocks of matter. An atom consists of three basic particles: (1) neutrons (neutral particles), (2) protons (positively charged particles), and (3) electrons (negatively charged particles). Neutrons and protons combine to form the positively charged nucleus which is the central part of the atom. The electrons revolve around the nucleus in different orbits. Atoms of different types are known as elements. Elements differ in the number of protons and electrons they have, but they have an equal number of each. When atoms of an element have a different number of neutrons, they are called isotopes of that element. Elements have many isotopes, and some of them may be unstable.

Radiation is energy transmitted in the form of waves or particles. There are two basic types of radiation: particulate radiation and electromagnetic radiation. Particulate radiation (alpha and beta radiation) has both mass and energy associated with it. Electromagnetic radiation is pure energy with no mass, such as x-rays and gamma rays. Radiation is produced when unstable isotopes undergo spontaneous change, known as radioactive disintegration or radioactive decay. The rate of decay is measured by how long it takes for half of the sample to decay. When an unstable isotope changes into a more stable form it may emit either an alpha particle or a beta particle. These reactions may or may not be associated with gamma radiation. The alpha and beta particles are generally referred to as ionizing radiation.

An alpha particle emits positively charged, highly energetic ionizing radiation that consists of two protons and two neutrons. Alpha particles are extremely limited in their ability to penetrate matter, and they can be stopped easily by a sheet of paper or by the outer layer of the skin. In air, they can travel only a few centimeters. Therefore, alpha particles outside the body do not cause any external radiation exposure. However, when the alpha particles are ingested or inhaled they dissipate all their energy in the living tissue, which results in radiation exposure.

A beta particle is an electron that is much lighter than an alpha particle. It can travel a longer distance in air than an alpha particle but can still be stopped by a thin sheet of aluminum foil. Low-energy beta emitters in general do not result in external radiation exposure, but high-energy beta emitters, when stopped by shielding, may generate Bremsstrahlung x-rays that may result in external radiation exposure. The intake of beta particles would result in internal radiation exposure.

X-rays and gamma rays are waves of pure energy that travel with the speed of light and are very penetrating; they require thick concrete or lead shielding to stop them. X-rays and gamma rays can result in both external and internal radiation exposure.

Neutrons lose energy through collisions with the nuclei of the atoms in their environment. They generally slow down to thermal or near thermal energies and are captured by nuclei of the absorbing material. Therefore, neutrons generally travel long distances in air or metallic components before they are absorbed. Radiation exposure occurs from gamma rays and alpha particles that are emitted when a neutron is captured in matter.

D.8.1.2 Conventional Quantities and Units

Following is the list of conventional terms used in the evaluation of radiological human health impacts.

- Absorbed dose: The energy imparted by ionizing radiation per unit mass of the irradiated material. The units of absorbed dose are the rad and the gray (Gy).
- Activity: The rate of disintegration (transformation) or decay of radioactive material. The units of radioactivity are the curie (Ci) and the Becquerel (Bq).
- Collective dose: The sum of the individual doses received in a given period of time by a specified population from exposure to a specified source of radiation.
- Committed dose equivalent: The dose equivalent to organs or tissues of reference that will be received from an intake of radioactive material by an individual during the 50-year period following the intake.
- Committed effective dose equivalent: The sum of the products of the weighting factors applicable to each of the body organs or tissues that are irradiated and the committed dose equivalent to these organs or tissues.
- Deep-dose equivalent: Applies to external whole-body exposure and is the dose equivalent at a tissue depth of 1 cm.
- Dose equivalent: The product of the absorbed dose in tissue, quality factor, and all other modifying factors at the location of interest. The units of dose equivalent are the rem and the sievert (Sv).
- Effective dose equivalent: The sum of the products of the dose equivalent to the organ or tissue and the weighting factors applicable to each of the body organs or tissues that are irradiated.
- External dose: That portion of the dose equivalent received from radiation sources outside the body.

- Internal dose: That portion of the dose equivalent received from radioactive material taken into the body.

- Nonstochastic effect: Health effects, the severity of which varies with the dose and for which a threshold is believed to exist. Radiation-induced cataract formation is an example of a nonstochastic effect (also called a deterministic effect).

- Organ dose: Dose received as a result of radiation energy absorbed in a specific organ.

- Occupational dose: Dose received by an individual in the course of employment in which the individual's assigned duties involve exposure to radiation or to radioactive material.

- Public dose: The dose received by a member of the public from exposure to radiation or radioactive material.

- Quality factor: The modifying factor (see Table D.8-1) that is used to derive the dose equivalent from the absorbed dose.

- Shallow-dose equivalent: Applies to external exposure of the skin or an extremity and is taken as the dose equivalent at a tissue depth of 0.007 cm averaged over an area of 1 cm^2.

- Stochastic effect: Health effects that occur randomly and for which the probability of the effect occurring, rather than its severity, is assumed to be a linear function of dose without threshold. Hereditary effects and cancer incidence are examples of stochastic effect.

Table D.8-1. Quality Factors and Absorbed Dose Equivalencies

Type of Radiation	Quality Factor	Absorbed Dose Equal to a Unit Dose Equivalent[(a)]
X-, gamma, or beta radiation	1	1
Alpha particles, multiple-charged particles, fission fragments, and heavy particles of unknown energy.	20	0.05
Neutrons of unknown energy	10	0.1
High energy protons	10	0.1

(a) Absorbed dose in rad equal to 1 rem or the absorbed dose in gray equal to sievert.
Source: 10 CFR Part 20

- Total body dose: Sum of the dose received from external exposure to the total body, gonads, active blood-forming organs, head and trunk, or lens of the eye and the dose due to the intake of radionuclides by inhalation and ingestion where a radioisotope is uniformly distributed throughout the body tissues rather than being concentrated in certain parts.

- Total effective dose equivalent: The sum of the deep-dose equivalent (for external exposure) and the committed effective dose equivalent (for internal exposure).

- Weighting factor: The fraction of the overall health risk, resulting from uniform whole body irradiation, attributable to a specific organ or tissue. Table D.8-2 lists organ dose weighting factors.

- Whole body dose: Same as total body dose.

D.8.1.3 Biological Effects of Radiation

Radiation interacts with the atoms that form the cells. There are two mechanisms by which radiation affects cells: direct action and indirect action. In a direct action, the radiation interacts directly with the atoms of the DNA molecule or some other component critical to the survival of the cell. Since the DNA molecules make up a small part of the cell, the probability of direct action is small. Because most of the cell is made up of water, there is a much higher probability that radiation would interact with water. In an indirect action, radiation interacts with water and breaks the bonds that hold the water molecule together and produces reactive free radicals that are chemically toxic and destroy the cell. The body has mechanisms to repair damage caused by radiation. Consequently, biological effects of radiation on living cells may result in three outcomes: (1) injured or damaged cells repair themselves, resulting in no residual damage; (2) cells die, much like millions of body cells do every day, being replaced through normal biological processes; or (3) cells incorrectly repair themselves, resulting in a biophysical change. Stochastic effects may occur when an irradiated cell is modified rather than killed. A modified cell may, after a prolonged delay, develop into a cancer.

Table D.8-2. Organ Dose Weighting Factors

Organ or Tissue	Weighting Factor
Gonads	0.25
Breast	0.15
Red bone marrow	0.12
Lung	0.12
Thyroid	0.03
Bone surfaces	0.03
Remainder	0.30[(a)]
Whole body	1.00[(b)]

(a) 0.30 results from 0.06 for each of five "remainder" organs (excluding the skin and the lens of the eye) that receive the highest doses.
(b) For the purpose of weighting the external whole body dose (for adding it to the internal dose), a single weighting factor of 1 has been specified.
Source: 10 CFR Part 20

The biological effects on the whole body from exposure to radiation depend on many factors, such as the type of radiation, total dose, time interval over which the dose is received, and part of the body that is exposed. Not all organs are equally sensitive to radiation. The blood-forming organs are most sensitive to radiation; muscle and nerve cells are relatively insensitive to radiation. There could be two types of radiation exposure: (1) a single accidental exposure to high doses of radiation for a short period of time (acute exposure), which may produce biological effects within a short time after exposure, and (2) long-term, low-level overexposure, commonly called continuous or chronic exposure. High doses of radiation can cause death. Other possible effects of a high radiation dose include erythema, dry desquamation, moist desquamation, hair loss, sterility, cataracts, and acute radiation syndromes. Low doses of radiation can cause genetic effects and carcinogenic effects.

D.8.1.4 Human Health Effects of Radiation

Radiation can cause a variety of health effects. The most significant of these are induced cancer fatalities. The National Research Council's Committee on the Biological Effects of Ionizing Radiation (BEIR) has prepared a series of reports on the health consequences of radiation exposure. In the 1996 GEIS (NRC 1996) the NRC staff summarized the risk estimates from different reports including BEIR-I, BEIR-III, and BEIR-V, the 1988 UNSCEAR (United Nations Scientific Committee on the Effects of Atomic Radiation) reports, and International Commission on Radiological Protection (ICRP) Publication 26 (ICRP 1977).

In 1991, the ICRP issued a complete set of new recommendations based on new biological information (ICRP 1991). Table D.8-3 provides the nominal probability coefficients for stochastic effects. ICRP estimated the probability of fatal cancer by using the data on the Japanese survivors of the Hiroshima and Nagasaki atomic bombs and their assessment by

Table D.8-3. Nominal Probability Coefficients for Stochastic Effects

	Probability Coefficients (10^{-4} rem^{-1})(a)			
Exposed Population	**Fatal Cancer**	**Nonfatal Cancer**	**Total Cancer**	**Severe Hereditary Effects**
Adult workers	4.0	0.8	4.8	0.8
Whole population	5.0	1.0	6.0	1.3

(a) Rounded values.
Source: ICRP 1991

BEIR and UNSCEAR committees. ICRP reviewed the available experimental data on dose-response relationships for radiation of low linear energy transfer (LET) and the effect of dose and dose rate on this relationship and concluded that the dose-response relationship is most probably linear quadratic for low LET radiations. The BEIR-V risk estimate (eight cancer fatalities among 10,000 people exposed to 10,000 person-rem) was based on a high dose. ICRP in its 1991 recommendations used a dose and dose rate effectiveness factor (DDREF) of 2 to convert the high-dose or high-dose-rate estimates of risk to low-dose or low-dose-rate estimates of risk. The estimates of severe hereditary effects were also based on the experimental data on genetic effects in animals, which were in assessments done by BEIR and UNSCEAR committees.

In 1993, the National Council on Radiation Protection and Measurement (NCRP) recommended that a lifetime risk of fatal cancer of 4×10^{-4}/(person-rem) be used for a worker population and similarly, a lifetime risk of 5×10^{-4}/(person-rem) be used for the general population. The NCRP also recommended a risk estimate of about 1×10^{-4}/(person-rem) for severe hereditary effects in the total population and a somewhat lower risk estimate for the worker population (NCRP 1993). These recommendations are similar to the ICRP recommendations on the lifetime risk of fatal cancer.

In 1999, the EPA issued Federal Guidance Report No. 13, which provides numerical factors for use in estimating the risk of cancer from low-level exposure to radionuclides. Risk coefficients are provided for the following modes of exposure to a given radionuclide: inhalation of air, ingestion of food, ingestion of tap water, external exposure from submersion in air, external exposure from the ground surface, and external exposure from soil contaminated to an infinite depth (EPA 1999). The risk coefficients are applicable to either chronic or acute exposure to a radionuclide.

In 2006, the National Research Council's Committee on the Biological Effects of Ionizing Radiation (BEIR) published its latest report BEIR-VII, *Health Risks from Exposure to Low Levels of Ionizing Radiation* (BEIR 2006). The committee had published its previous report on the same topic, BEIR-V, in 1990 (BEIR 1990).

Three major changes have occurred after the BEIR V report was published. First, an additional 12 years of follow-up medical data were available. Second, cancer incidence data for the cohort were available (for BEIR V, only mortality data were available). The impact of these two developments has reduced the uncertainty in the assessment of cancer risk among the atomic bomb survivors. Third, the dosimetry system used to assign radiation exposure to the atomic bomb survivors was replaced with an improved dosimetry system. These changes have improved the understanding of the health risks associated with radiation exposure (NRC 2005).

In estimating the cancer risk, the committee used the Hiroshima and Nagasaki atomic bomb survival data for the period 1958–1998 and a dose and dose rate effectiveness factor of 1.5 was used. Table D.8-4 lists the recommended risk coefficients for cancer incidence and fatality. Table D.8-4 shows the estimated cancer cases and deaths in the U.S. population that would be expected to result if each individual in a population of 100,000 was exposed to a single dose of 10 rad. It also shows the number that would be expected in the absence of radiation. The 95 percent confidence intervals are also shown.

The BEIR VII committee's preferred estimate of lifetime attributable risk for solid cancer incidence and mortality (Table D.8-4) suggests that females are more sensitive than males to radiation exposure at 10 rem, a level that is 100 times the NRC's radiation protection standards specified in 10 CFR Part 20. The BEIR VII committee's preferred estimate of lifetime attributable risk for leukemia cancer incidence and mortality (Table D.8-4), moreover, suggests that males are more sensitive than females. The BEIR VII committee uses the 95 percent confidence intervals associated with estimated lifetime cancer risk for males and females that suggest that the apparent gender difference may not be statistically significant.

Table D.8-5 compares the BEIR VII risk estimates for whole population with estimates recommended by BEIR V, ICRP, EPA, and UNSCEAR in recent years. The overall difference in the risk estimates recommended by different organizations is statistically insignificant. In this regard, the BEIR VII report states: "in general the magnitude of estimated risks for total cancer mortality or leukemia has not changed greatly from estimates in past reports such as BEIR V and recent reports of the United Nations Scientific Committee on the Effects of Atomic

Table D.8-4. Estimates of Lifetime Attributable Risk of Incidence and Mortality for All Solid Cancers and for Leukemia in the BEIR VII Report[(a)]

	All Solid Cancers		**Leukemia**	
Category	**Males**	**Females**	**Males**	**Females**
Excess cases (including nonfatal cases) from exposure to 10 rad	800 (400–1,600)	1300 (690–2,500)	100 (30–300)	70 (20–250)
Number of cases in the absence of exposure	45,500	36,900	830	590
Excess deaths from exposure to 10 rad	410 (200–830)	610 (300–1,200)	70 (20–220)	50 (10–190)
Number of deaths in the absence of exposure	22,100	17,500	710	530

(a) Number of cases or deaths per 100,000 exposed persons with 95 percent subjective confidence intervals shown in parentheses.

Source: BEIR 2006

Table D.8-5. Comparison of BEIR VII Lifetime Cancer Mortality Estimates with Those from Other Reports

Cancer Category	BEIR V[(a)] (1990)	ICRP[(b)] (1991)	EPA[(b)] (1999)	UNSCEAR[(c)] (2000)	BEIR VII[(d)] (2006)
Leukemia[(e)]	50	56	50	–[(f)]	61
All cancer except leukemia	460	450	520	–	–
All solid cancers (sum)	–	–	–	520	510

NOTE: Excess deaths for population of 100,000 of all ages and both sexes exposed to 10 rad.
(a) Average of estimates for males and females. The values show the results that would be obtained if the DDREF of 1.5, used by the BEIR VII committee, had been employed.
(b) Except for the EPA breast and thyroid cancer estimates, the solid cancer estimates are linear estimates reduced by a DDREF of 2.
(c) Average of estimates for males and females. The estimate is a combined estimate (using the same weights as used by the BEIR VII committee applied on a logarithmic scale) reduced by a DDREF of 1.5.
(d) Average of the committee's preferred estimates for males and females.
(e) Estimates based on a linear-quadratic model.
(f) Not reported.
Source: BEIR 2006

Radiation (UNSCEAR) and the International Commission on Radiological Protection (ICRP). New data and analyses have reduced sampling uncertainty, but uncertainties related to estimating risk for exposure at low doses and dose rates and transporting risks from Japanese A-bomb survivors to the U.S. population remain large. Uncertainties in estimating risks of site specific cancers are especially large."

If the total fatal cancer risk is the sum of cancer deaths from all solid cancers and leukemia, then the fatal cancer risk coefficient for the general public would be 6×10^{-4}/person-rem (see Table D.8-5). The fatal cancer risk for the general public based on ICRP is 5×10^{-4}/person-rem (Table 3.9-20). There is a difference of approximately 20 percent in the fatal cancer risk coefficient based on the ICRP recommendation and the BEIR-VII report. The difference of 20 percent is within the margin of uncertainty associated with these estimates.

D.8.1.5 Methodology for Estimating Radiological Impacts

Radiological exposures from nuclear power plants include offsite doses to members of the public and onsite doses to the workforce. Nuclear power plants must be licensed by the NRC and comply with NRC regulations and conditions specified in the license. The licensees are required to comply with 10 CFR Part 20, Subpart C, "Occupational Dose Limits for Adults," and 10 CFR Part 20, Subpart D, "Radiation Dose Limits for Individual Members of the Public" (see Section 3.9.1.1).

D.8.1.5.1 Methodology for Estimating Worker Doses

Plant workers conducting activities involving radioactively contaminated systems or working in radiation areas can be exposed to radiation. Most of the occupational radiation dose to nuclear plant workers results from external radiation exposure rather than internal exposure from inhaled or ingested radioactive materials. Workers also receive radiation exposure during the storage and handling of radioactive waste and during the inspection of stored radioactive waste. However, these sources of exposure are small when compared with other sources of exposure at operating nuclear plants.

Individual occupational doses are measured by NRC licensees as required by the basic NRC radiation protection standard, 10 CFR Part 20 (Section 3.9.1.1). This standard includes requirements for summing internal and external dose equivalents to yield the total effective dose equivalent (TEDE).

Worker doses from external exposure at a nuclear power plant are measured by using either a thermoluminescence dosimeter (TLD) or a film badge. Workers at nuclear plants, in addition to wearing these, wear direct-reading dosimeters (electronic dosimeters) in order to monitor occupational doses related to specific jobs. A TLD may be a Teflon disc impregnated with lithium fluoride sealed in a polyethylene envelope. The TLD is the most widely used personal monitor for gamma radiation and charged particles. Direct-reading dosimeters are useful in situations where there is the potential for sudden or large increases in exposure rate.

The potential external exposure for workers involved in radioactive waste management will likely result from gamma and beta radiation, and the use of the external monitoring devices discussed above is necessary. Internal dosimetry is used when there is a potential that the body may have taken in radioactive material. There are two methods to calculate the committed dose equivalent: (1) measurement of the airborne concentration and the time a worker spends in that area and (2) urinalysis and monitoring of feces or blood. At nuclear power plants, method 1 is generally used. However, for complex situations the mathematical models of the radionuclide's retention and excretion are generally used, as are measurements of the radioactive material content in the excreta, to estimate the doses. Bioassay techniques, such as urinalysis, provide a screening tool to maintain and verify operational radiation protection and control.

For this GEIS revision, worker dose information was obtained from the 38th annual report titled *Occupational Radiation Exposure at Commercial Nuclear Power Reactors and Other Facilities 2005* (Burrows and Hagemeyers 2006). This report summarizes the occupational exposure data maintained by the NRC's Radiation Exposure Information and Reporting System (REIRS). The licensees submit radiation exposure records for each monitored individual.

D.8.1.5.2 Methodology for Estimating Public Doses

Commercial nuclear power plants, under normal operations, release small amounts of radioactive materials to the environment. The effluent releases (gaseous and liquid) result in radiation doses to humans. Nuclear power plant licensees must comply with Federal regulations (e.g., 10 CFR Part 20, Appendix I to 10 CFR Part 50, 10 CFR Part 50.36a, and 40 CFR Part 190) and conditions specified in the operating license (see Section 3.9.1.1). Appendix I to 10 CFR Part 50 provides numerical values for radioactive effluent design objectives. In addition, each plant license contains technical specification requirements for controlling and limiting the discharge of radioactive gaseous and liquid effluents.

Potential environmental pathways through which persons may be exposed to radiation originating in a nuclear power reactor include atmospheric and aquatic pathways. Radioactive materials released under controlled conditions include fission products and activation products. Fission product releases consist primarily of the noble gases and some of the more volatile materials like tritium, isotopes of iodine, and cesium. These materials are monitored carefully before release to determine whether the limits on releases can be met. Releases into aquatic systems are similarly monitored. When an individual is exposed through one of these pathways, the dose is determined in part by the exposure time, the amount of material ingested or inhaled, and in part by the amount of time that the radioactivity inhaled or ingested is retained in the individual's body. The major exposure pathways include the following:

- Inhaling contaminated air,
- Drinking milk or eating meat from animals that graze on open pasture on which radioactive contamination may be deposited,
- Eating vegetables grown near the site, and
- Drinking (untreated) water or eating fish caught near the point of discharge of liquid effluents.

Other less important exposure pathways include external irradiation from surface deposition; consumption of animals that drink water that may contain liquid effluents; consumption of crops grown near the site using irrigation water that may contain liquid effluents; shoreline, boating, and swimming activities; and direct offsite irradiation from radiation coming from the plant.

To implement Appendix I of 10 CFR Part 50, the NRC has developed a series of Regulatory Guides that present methods it finds acceptable for calculating effluent releases, the dispersion of effluent in the atmosphere and different water bodies, and the associated radiation doses. In general, licensees follow the guides developed by the NRC staff to calculate public doses.

Liquid effluent from a nuclear power plant may be released into a variety of surface water bodies (e.g., rivers, lakes, reservoirs, cooling ponds, estuaries, and open coastal waters). The released liquid effluent is dispersed by turbulent mixing and by stream flow in rivers, by tidal or nontidal coastal currents in estuaries and coastal waters, and by internal circulation or flow-through in lakes, reservoirs, and cooling ponds. Many parameters (e.g., direction and speed of the flow of currents in the receiving water bodies; size, geometry, and bottom topography of the water body) influence dispersion and dilution. Revision 1 of Regulatory Guide 1.113 (NRC 1977a) describes calculational models for estimating the aquatic dispersion of routine or accidental releases of radioactive material from a nuclear power plant to a surface water body.

Gaseous effluents from nuclear power plants are mostly released through tall stacks or vents near the top of buildings. In some cases, releases could occur near ground level; an example is when auxiliary equipment or a component such as a waste storage tank is housed outside the buildings. Effluent concentrations at downwind locations depend on many parameters (e.g., the initial release height, size and shape of the release point, initial vertical velocity of the effluent, heat content of the effluent, ambient wind speed and temperature, atmospheric stability, and effluent removal mechanisms). Geographic features such as hills, valleys, and large bodies of water greatly influence dispersion and airflow patterns. Revision 1 of Regulatory Guide 1.111 (NRC 1977b) describes basic features of the calculational models and assumptions used to estimate the atmospheric transport and dispersion of gaseous effluents in routine releases from nuclear power plants.

Revision 1 of Regulatory Guide 1.109 (NRC 1977c) provides methods for calculating radiation doses to the public. Appendix A of the regulatory guide describes methods for calculating doses from liquid effluent pathways. The appendix includes the method for calculating doses from potable water, aquatic food, shoreline deposits, and foods grown on land with contaminated water. Appendix B of the regulatory guide describes models and assumptions for calculating doses from noble gases discharged to the atmosphere. It includes the annual gamma and beta air dose calculations and the annual total body and skin dose calculations from noble gas effluents. Appendix C of the regulatory guide provides models and assumptions for calculating doses from radioiodines and other radionuclides released in the atmosphere. It includes the annual external dose calculation from direct exposure to radioactivity deposited on the ground surface, annual dose from inhalation of radionuclides in air, calculation of the radionuclide concentration in food from airborne activity, and calculation of the annual dose from contaminated foods. Appendix D of the regulatory guide provides models for calculating population doses from nuclear power plant effluents.

Radiation doses to the public are calculated in two ways. The first calculation is for dose to the maximally exposed person (that is, the real or hypothetical individual potentially subject to maximum exposure). The second is for doses to the average individual and population. Doses are calculated by using site-specific data when available. For those cases in which site-specific data are not readily available, conservative (overestimating) assumptions are used to estimate doses to the public. For calculating the dose, Regulatory Guide 1.109 divides the population into four age groups: infants (0 to 1 year), children (1 to 11 years), teenagers (11 to 17 years), and adults (17 years and older). Doses are calculated for the maximum exposed individual from these four age groups and compared with the design objectives (Table 3.9-2). Regulatory Guide 1.109 includes the dose factors for these four age groups.

Every year licensees submit two reports to the NRC: an annual radiological environmental monitoring report and an annual radioactive effluent release report. For this GEIS update, public doses from gaseous and liquid effluent releases were obtained from a series of annual radioactive effluent release reports.

D.8.2 Chemical Hazards

In nuclear power plants, chemical effects could result from discharges of chlorine or other biocides, small-volume discharges of sanitary and other liquid wastes, chemical spills, and heavy metals leached from cooling system piping and condenser tubing. Although information was provided about certain types of chemicals used at nuclear power plants, chemical hazards were not specifically addressed in the 1996 GEIS, but the human health impacts of chemicals are included in this GEIS update (Section 3.9.2). Impacts of chemical discharges on human health are considered to be SMALL if the discharges to water bodies are within effluent limits designed to ensure the protection of water quality. The methodology for assessing effects on water quality and aquatic biota are covered in other parts of this appendix. Human health impacts from chemicals were assessed on the basis of information provided in the 1996 GEIS, published literature, publically available SEISs, and the decommissioning GEIS (NRC 2002).

D.8.3 Microbiological Hazards

Some microorganisms associated with cooling towers and the thermal discharges associated with nuclear power plants can have deleterious impacts on the health of plant workers and the public. The potential for adverse health effects on workers at nuclear power plants as a result of the enhancement of microorganisms is an issue for plants that use cooling towers. The potential for adverse health effects on the public from thermally enhanced microorganisms is an issue for nuclear plants with once-through cooling systems that use cooling ponds, lakes, or canals and that discharge to small rivers. These issues were evaluated by reviewing the information in the 1996 GEIS and published literature on organisms that could be enhanced by plant operation. All published SEISs were also reviewed for new and significant information pertaining to microbiological issues.

D.8.4 Electromagnetic Fields

Nuclear power plants have power transmission systems associated with them that consist of switching stations (or substations) located on the plant site and transmission lines located primarily offsite. Electric and magnetic fields, collectively referred to as the electromagnetic field (EMF), are produced by operating transmission lines. The issue of potential chronic effects from exposure to EMF surrounding transmission lines was evaluated by reviewing the relevant literature.

D.8.5 Other Hazards

Nuclear power plants are industrial facilities that have many of the typical occupational hazards found at any other electric power generation facility. Workers at or around nuclear power plants would be involved in some maintenance activities, electrical work, electric power line maintenance, and repair work and subject to potentially hazardous physical conditions (excessive heat, cold, pressure, etc.). The human health impact from occupational hazards was not discussed in the 1996 GEIS but is considered in this GEIS update (Section 3.9.5). The occupational hazards were evaluated by comparing the rate of fatal injuries and nonfatal occupational injuries and illnesses in the utility sector with the rate in all industries combined.

The workers and general public located at or around nuclear power plants and along the transmission lines are exposed to the potential for acute electrical shock from transmission lines. The shock hazard was evaluated by referral to the National Electric Safety Code (NESC).

D.9 Waste Management and Pollution Prevention

D.9.1 Description of Affected Resources and Region of Influence

Similar to most industrial facilities, nuclear power plants and other fuel-cycle facilities generate waste during their operation. The waste materials are often shipped offsite by truck, train, or in some cases by barge either for disposal or for processing. The wastes that are sent to a processing facility may be reused or recycled or they may be sent to a disposal facility after processing. The processing and handling that occur at the site of generation, including any packaging and loading of the wastes onto conveyance vehicles for shipment offsite, are considered part of the normal operations at that site, and the impacts associated with them are assessed as part of the normal operational impacts. Impacts associated with transportation and offsite processing and disposal are considered under the waste management impacts.

The primary resource that is affected by the disposal of waste materials is the land that is used for disposal. This land is assumed to be an irreversibly and irretrievably committed resource (see Section 4.4.3). The resources that are affected during processing and disposal of the wastes are similar to the resources affected during operation of any nuclear fuel cycle facility, including the nuclear power plants. As discussed in Chapter 4, these resources include land use and visual resources, air quality and noise, geology and soils, hydrology, ecology, cultural resources, socioeconomics, human health and safety, and environmental justice. During transportation, the main resources affected are human health and safety, air quality and noise, and socioeconomics. The impact assessment methodologies and the regions of influence for these resource areas are covered in other parts of this appendix.

D.9.2 Description of Impact Assessment

Historical data and experience were used to estimate the characteristics and quantities of wastes generated at nuclear power plants. These values are discussed in the main body of this document under waste management sections (see for example Sections 3.11, 4.1.1.10, 4.1.3.10, and 4.1.4). Table S-3 in 10 CFR 51.51(b) was the main source for waste generation numbers at other nuclear fuel cycle facilities. The assessment of impacts associated with transportation of waste materials to and from a nuclear power plant relied on the information provided in Table S-4 of 10 CFR 51.52, whereas the impacts of transportation among other fuel-cycle facilities were addressed as part of Table S-3 as discussed Section 4.1.4. The impacts at the offsite processing and disposal facilities are not explicitly evaluated in this document because each of these facilities would be operated pursuant to a permit or license issued by either a Federal or State agency. The impacts at those facilities would be addressed as part of the permitting or licensing process for those facilities. All operations including disposal activities at the disposal facilities would be within the bounds of analyses conducted to obtain the facility's permit or license. For example, the waste shipped to the disposal facility would have to meet that facility's waste acceptance criteria.

The issues associated with the availability of disposal facilities for low-level waste (LLW) are discussed in Section 4.1.1.10. Section 4.1.1.10 also discusses the onsite storage of spent nuclear fuel during the licensing term of a reactor. For all other waste types, it is assumed that permitted processing and/or disposal facilities will be available when needed. Historical evidence suggests that this assumption is valid.

Pollution prevention and waste minimization practices generally employed at the nuclear power plant sites are discussed in Section 3.11. These practices are based on the requirements placed on the licensees by the NRC, EPA, or other Federal or State agencies and the licensee's own efforts to minimize the emissions to the environment and minimize the quantities of wastes generated or sent offsite for treatment or disposal.

D.10 Alternatives

D.10.1 Identification and Evaluation of Replacement Power Alternatives

To ensure that the analysis of replacement power alternatives focused only on realistic options, data published by the DOE's Energy Information Administration (EIA) were used to identify the current and projected contributions made to the commercial electric power sector by various fossil fuel and renewable energy technologies. Federal and State regulations, as well as the Internet Web sites of Federal and State regulatory agencies and State coalitions were reviewed to identify current and anticipated environmental externalities that would most likely also influence alternative energy technology selections. As a result of these reviews, twelve fossil fuel technologies and eight renewable energy technologies were identified, together with a nuclear energy alternative, as likely replacements for a retiring nuclear reactor. In addition, demand-side management (DSM) and power purchases also were identified for consideration.

The environmental consequence analyses for those fossil fuel, nuclear, and renewable energy technologies selected as likely alternatives were based on data from a variety of sources. Engineering and environmental performance data for fossil fuel technologies were obtained from reports published by DOE's National Energy Technology Laboratory (NETL) and the EPA. Published environmental impact statements (EISs), regulatory guidance, and early site permit applications provided the basis for the environmental consequence analysis of the nuclear energy alternative. Reports and technology overviews published by DOE's Office of Energy Efficiency and Renewable Energy (EERE) and the National Renewable Energy Laboratory (NREL) served as the principal sources of data for environmental impacts of the selected renewable energy technologies. Resource maps developed by NREL were also used to show the geographic relationships between existing commercial nuclear power facilities and readily accessible renewable energy resources of sufficient size and quality to support utility-scale power production. Additional data regarding the environmental consequences of renewable energy technologies were obtained from EISs published by Federal and State agencies and from other sources within the open literature. Impact analyses for DSM and power purchases were supported by data from the EIA and the Federal Energy Regulatory Commission (FERC).

D.10.2 Supporting Information

Schematic diagrams of fossil energy technologies (Figures D.10-1 to D.10-12) and renewable energy technologies (Figures D.10-13 to D.10-15; D.10-19 to D.10-21; and D.10-23 to D.10-25) are presented in this section to aid in an understanding of the operational components of different energy alternatives. Many of the renewable energy technologies are not equally viable in all parts of the country because of the uneven distribution of the underlying energy source. To illustrate availability of renewable energy alternatives, resource distribution maps are also provided (Figures D.10-16 to D.10-18; D.10-22; D.10-26).

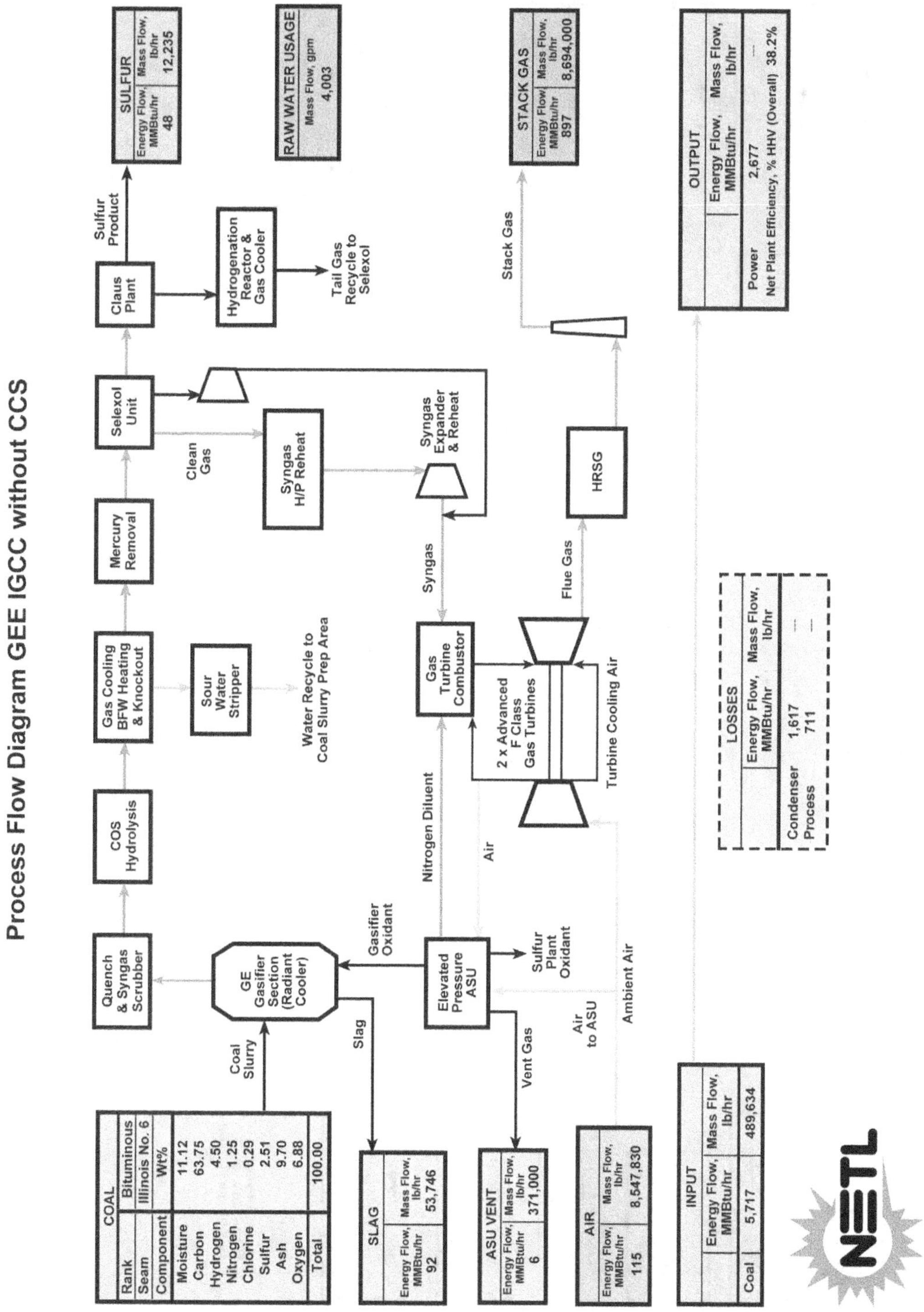

Figure D.10-1. Integrated Gasification Combined Cycle (IGCC) Coal Power Plant with GE Gasifier without CO_2 Capture (NETL 2007)

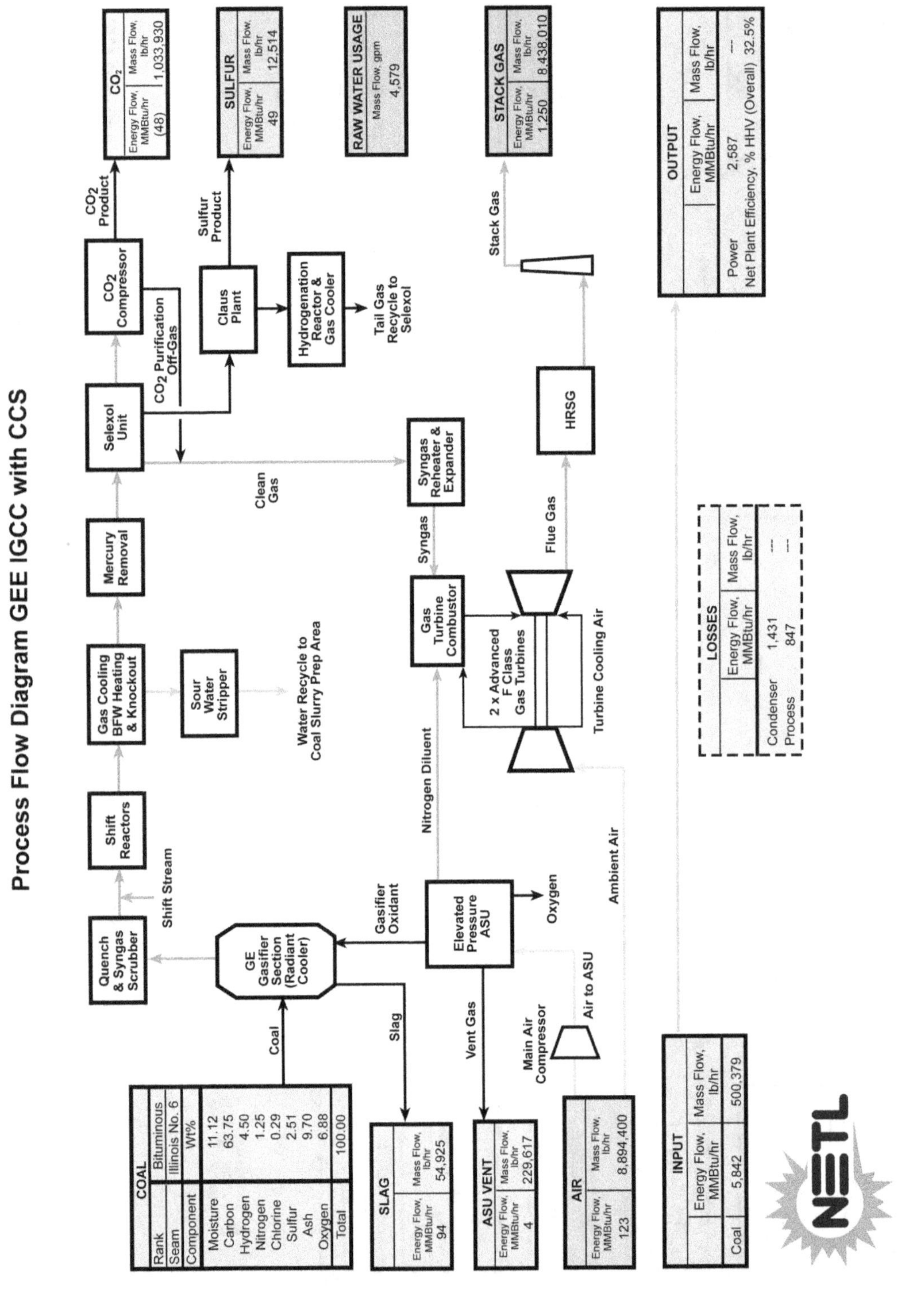

Figure D.10-2. IGCC Coal Power Plant with GE Gasifier with CO_2 Capture (NETL 2007)

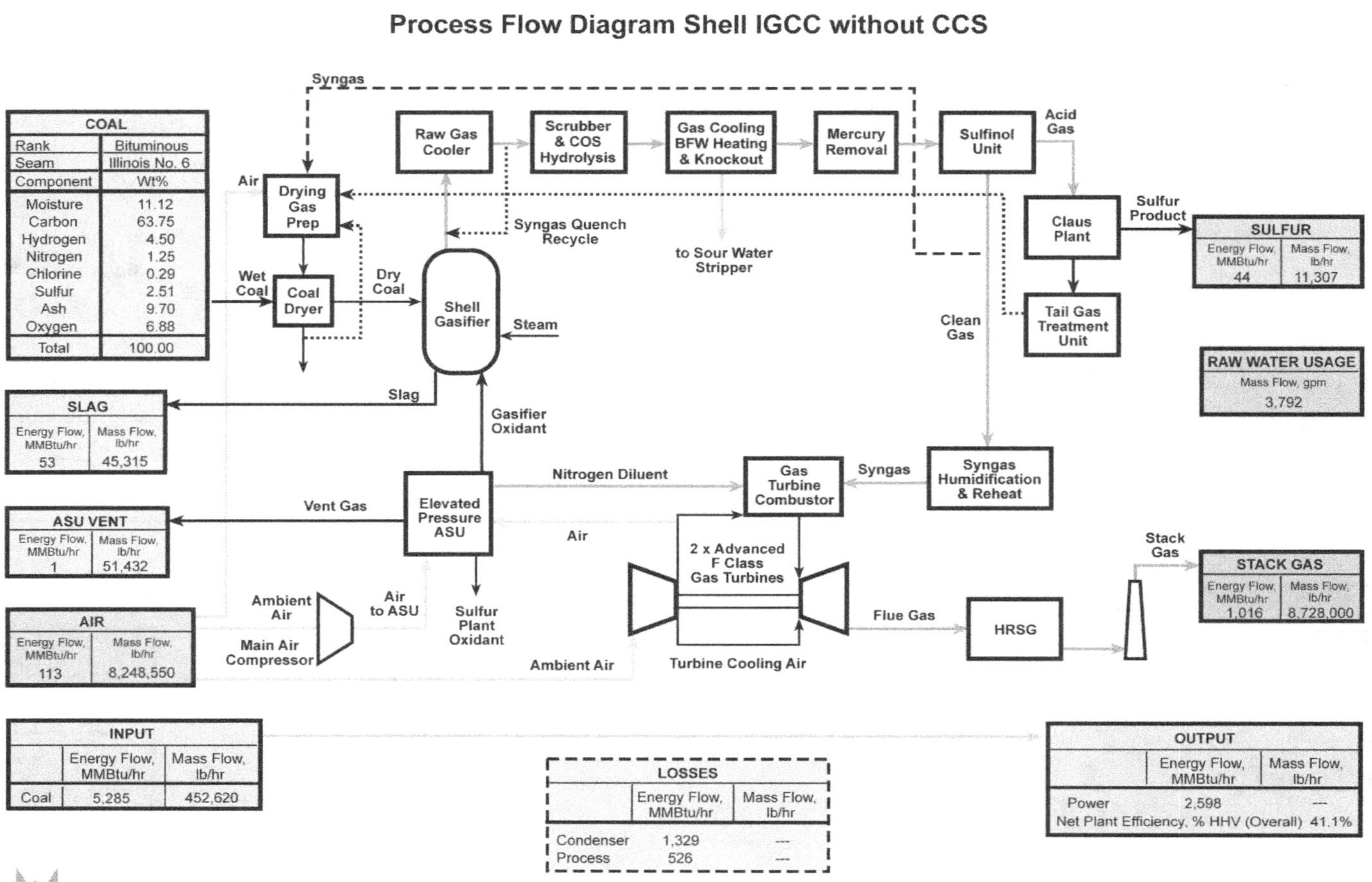

Figure D.10-3. IGCC Coal Power Plant with Shell Gasifier without CO_2 Capture (NETL 2007)

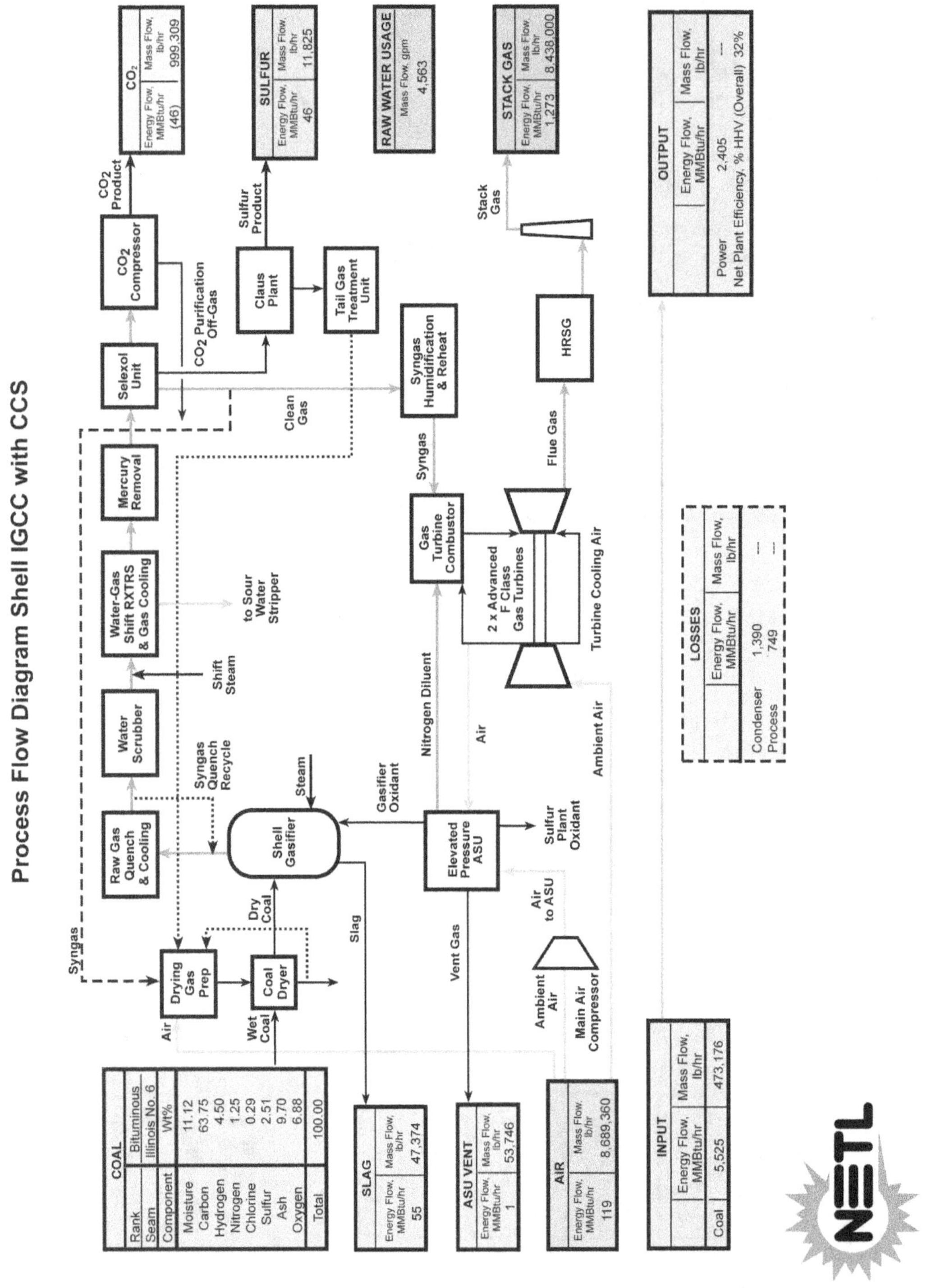

Figure D.10-4. IGCC Coal Power Plant with Shell Gasifier with CO_2 Capture (NETL 2007)

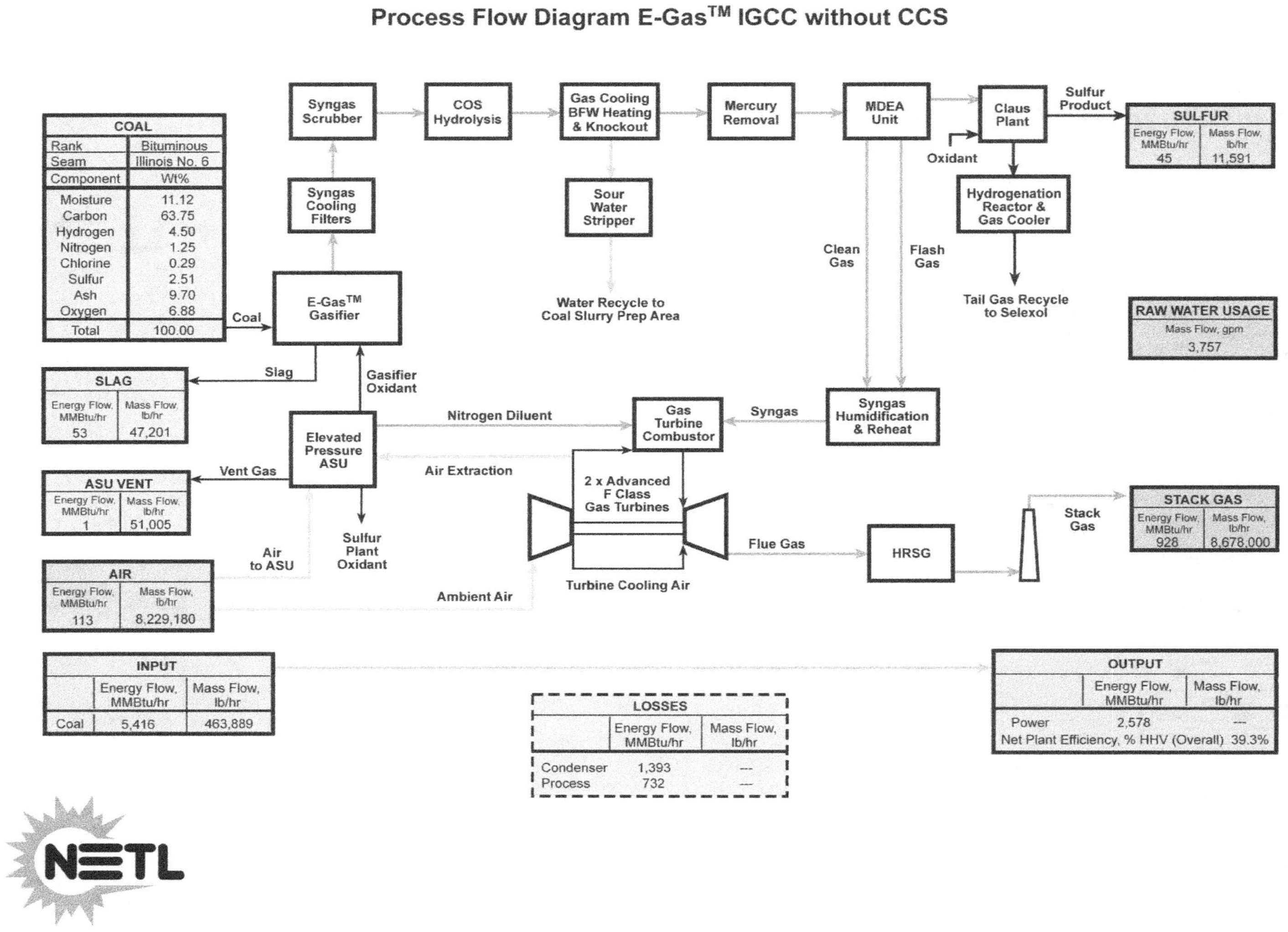

Figure D.10-5. IGCC Coal Power Plant with Conoco-Phillips (E-Gas^{TM}) Gasifier without CO_2 Capture (NETL 2007)

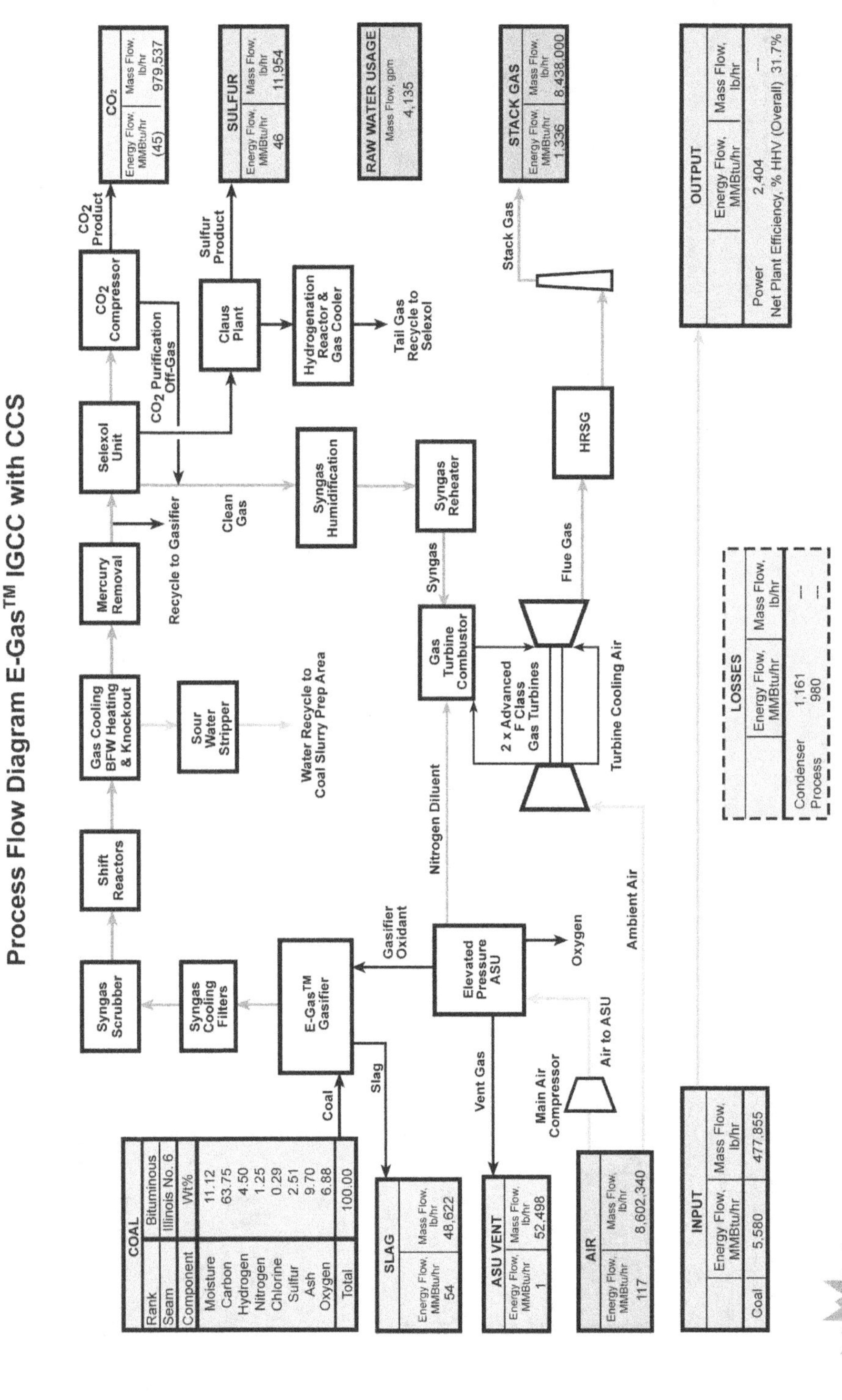

Figure D.10-6. IGCC Coal Power Plant with Conoco-Phillips (E-Gas™) Gasifier with CO_2 Capture (NETL 2007)

Process Flow Diagram Subcritical PC without CCS

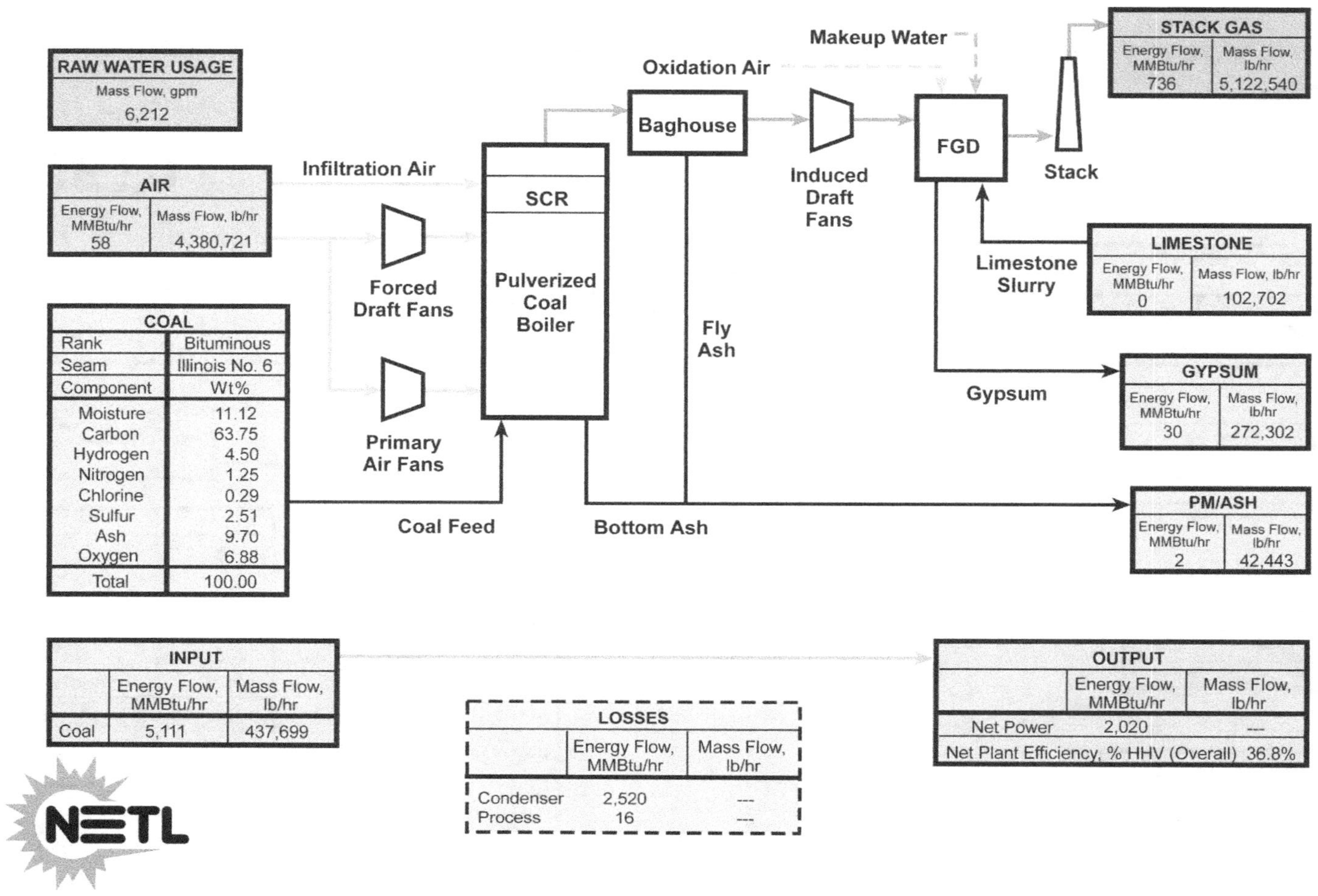

Figure D.10-7. Subcritical Pulverized Coal Power Plant without CO_2 Capture (NETL 2007)

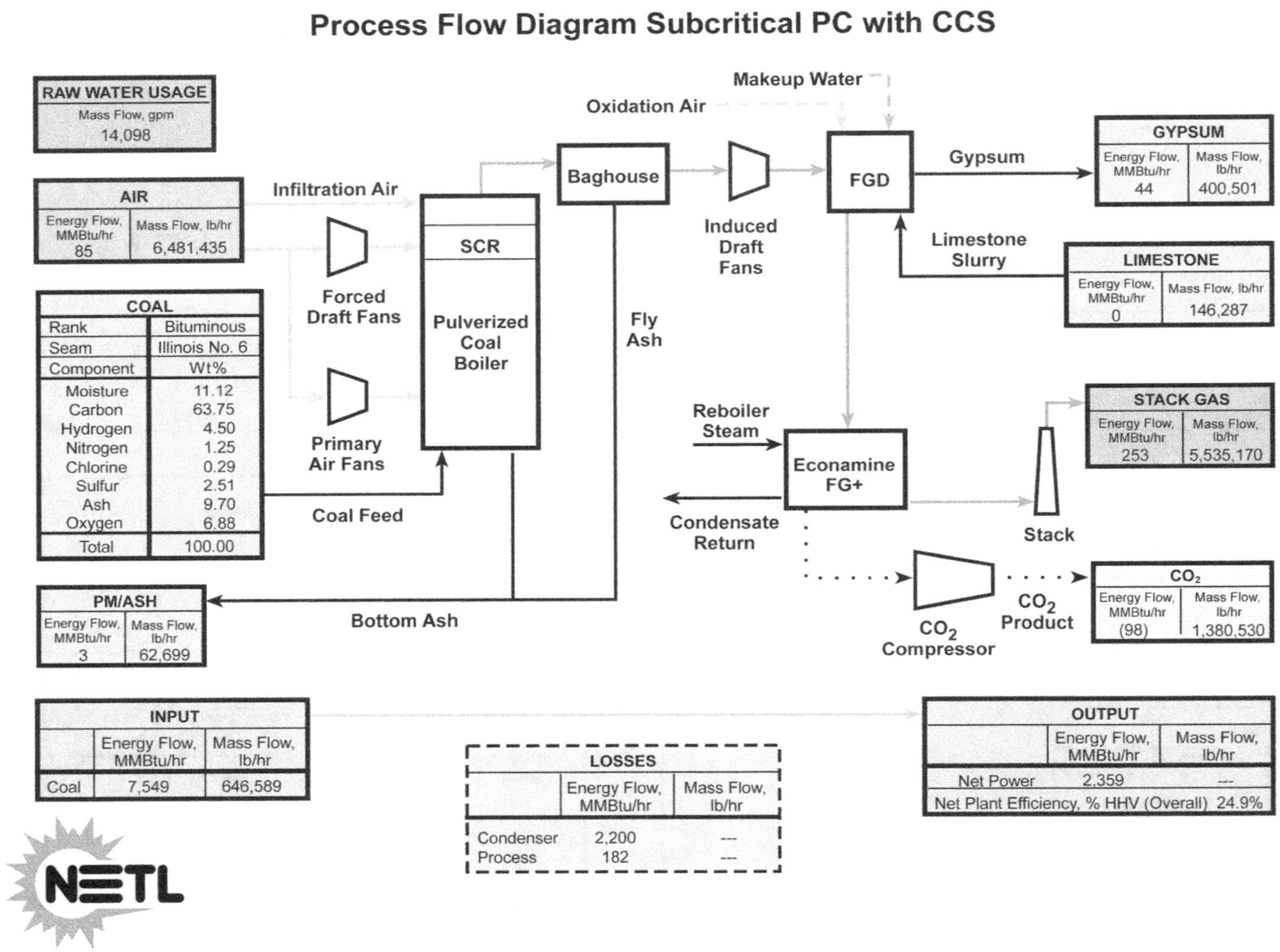

Figure D.10-8. Subcritical Pulverized Coal Power Plant with CO_2 Capture (NETL 2007)

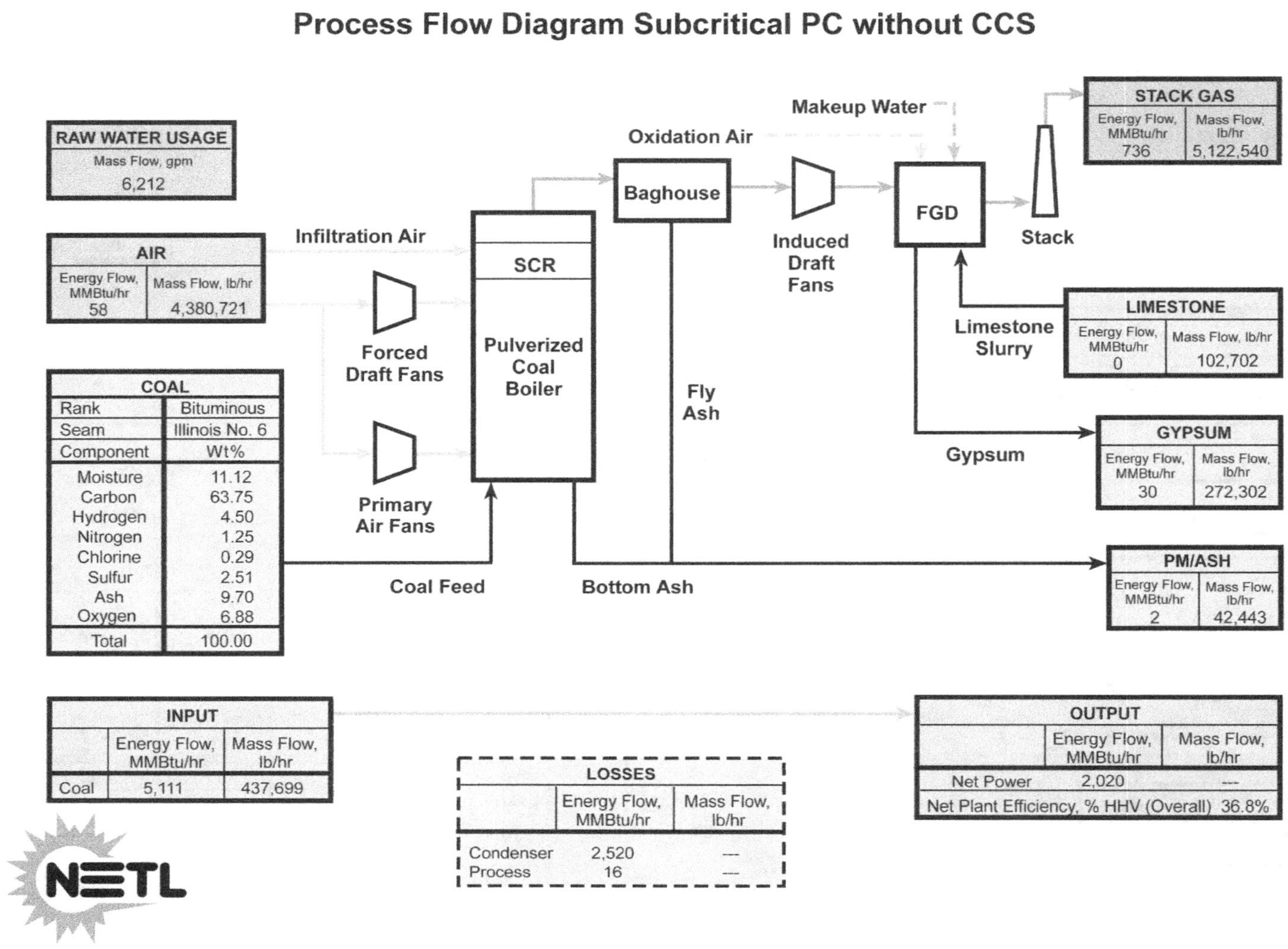

Figure D.10-9. Supercritical Pulverized Coal Power Plant without CO_2 Capture (NETL 2007)

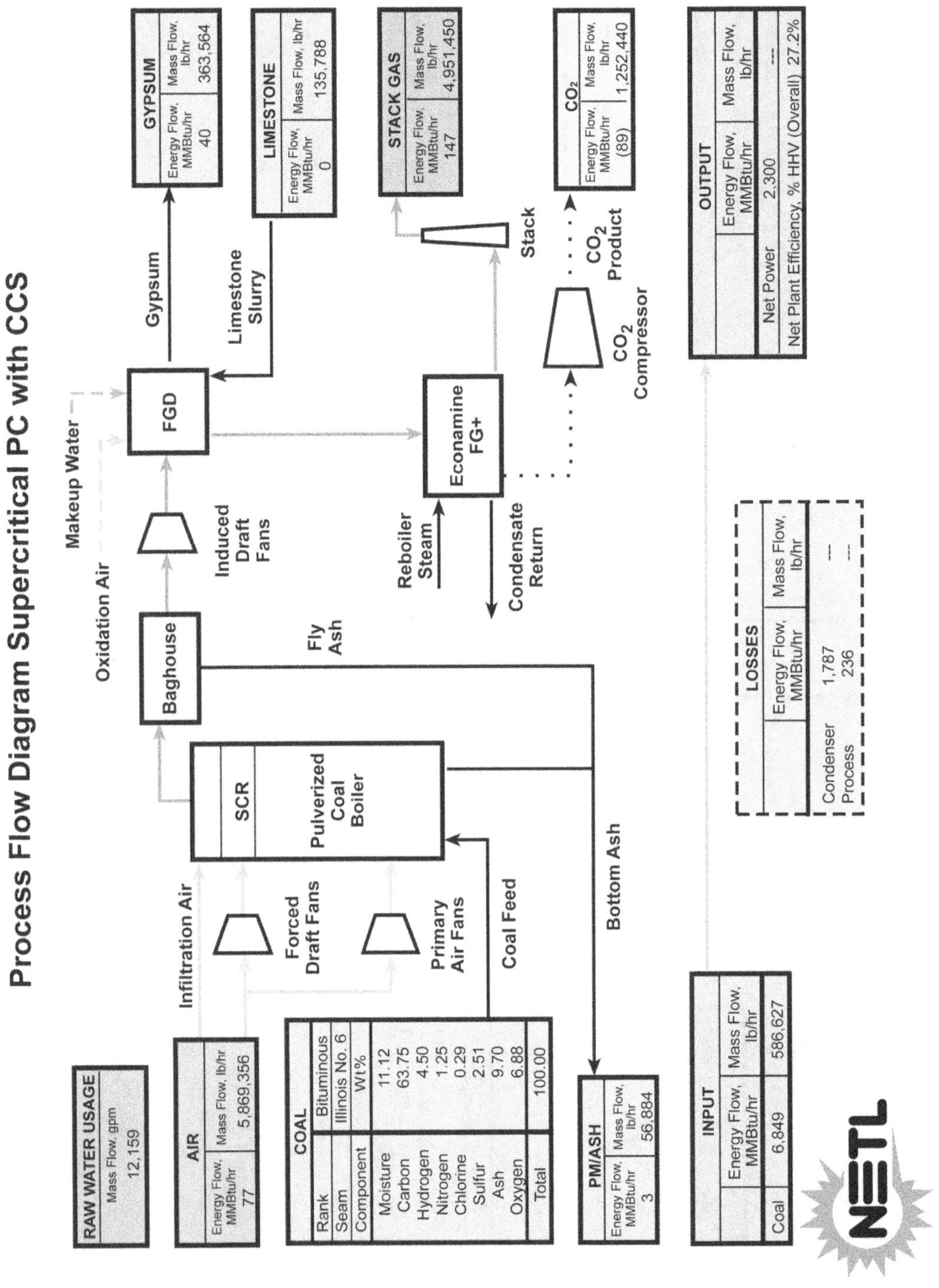

Figure D.10-10. Supercritical Pulverized Coal Power Plant with CO_2 Capture (NETL 2007)

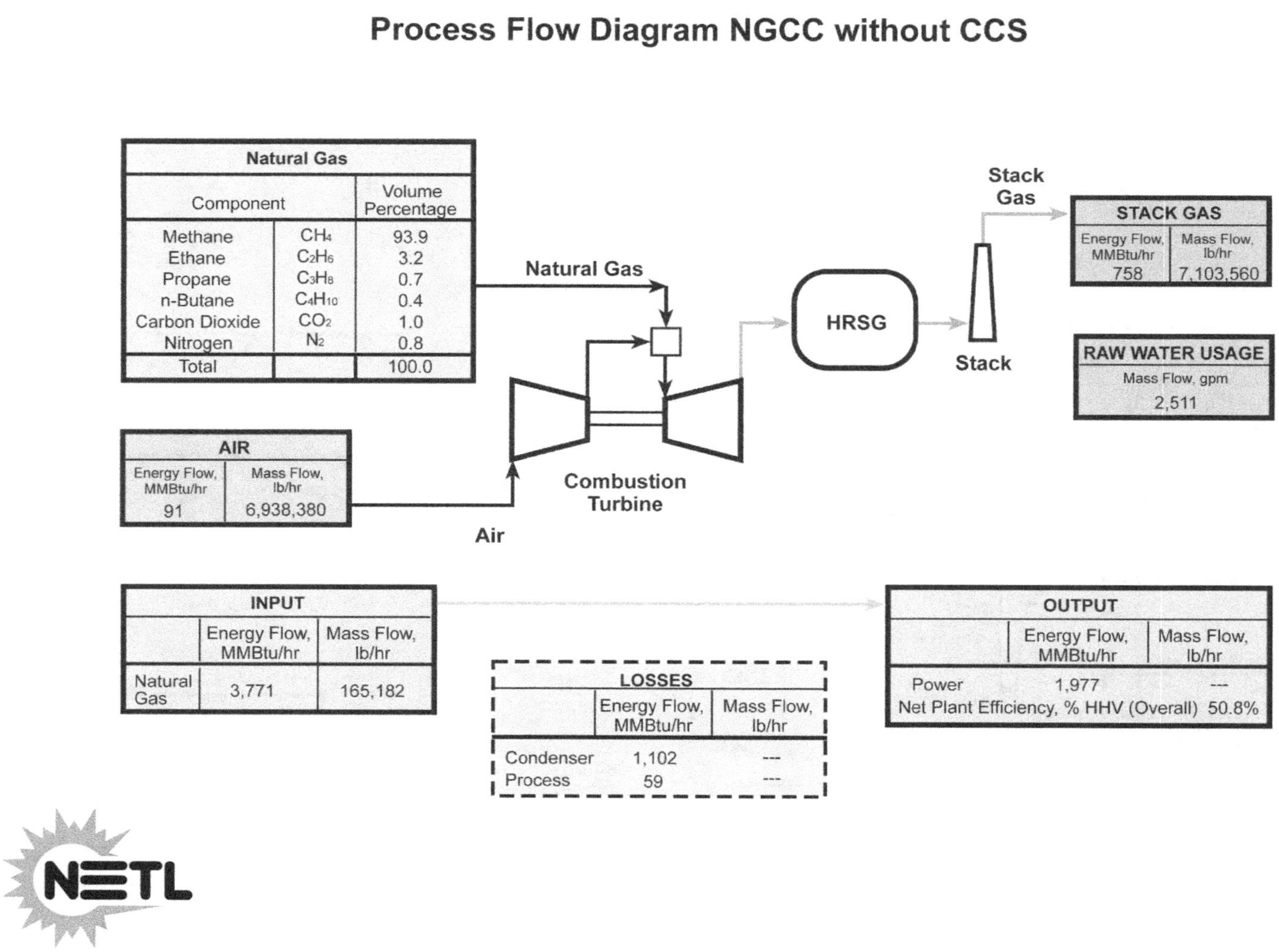

Figure D.10-11. Natural Gas IGCC Power Plant without CO_2 Capture (NETL 2007)

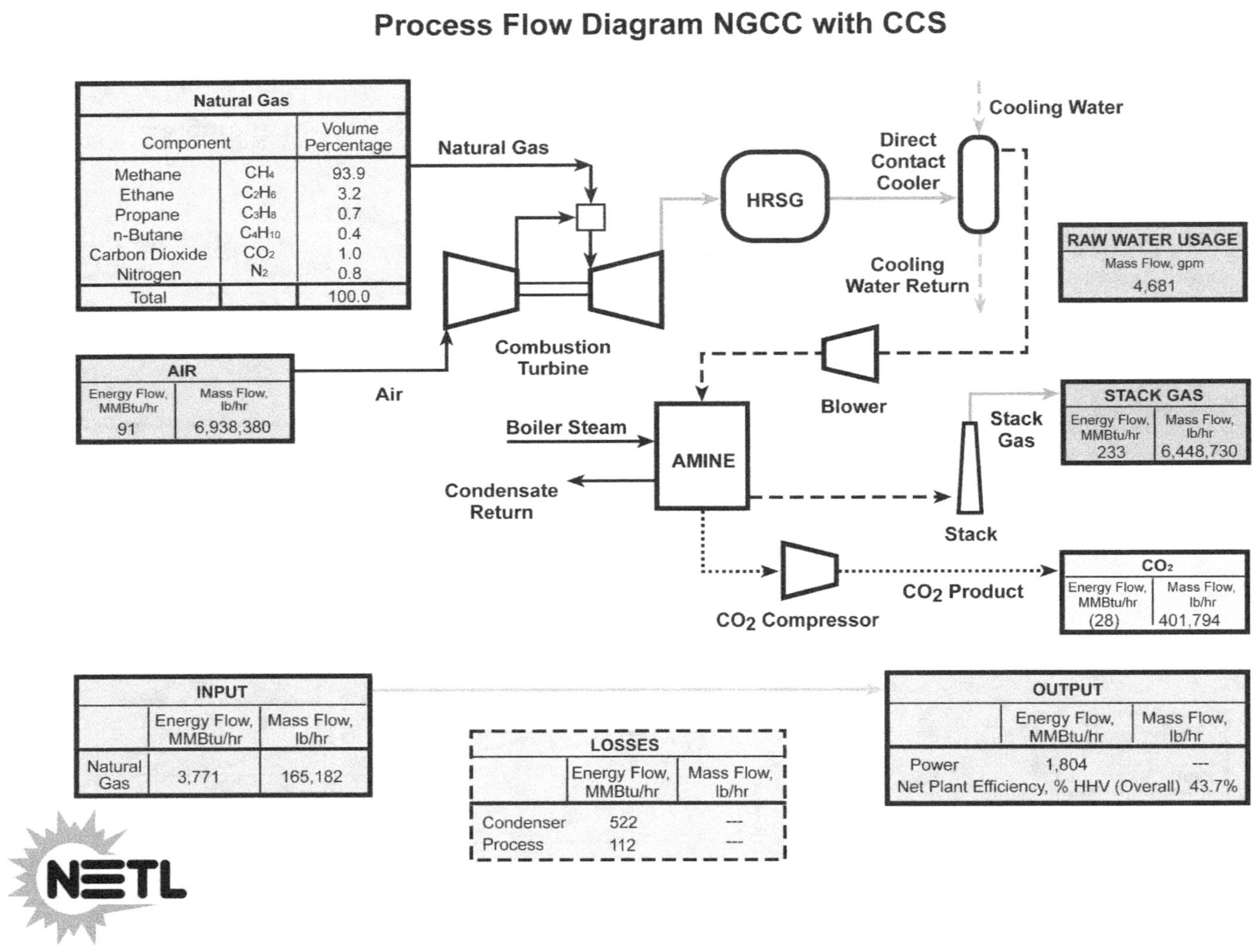

Figure D.10-12. Natural Gas IGCC Power Plant with CO_2 Capture (NETL 2007)

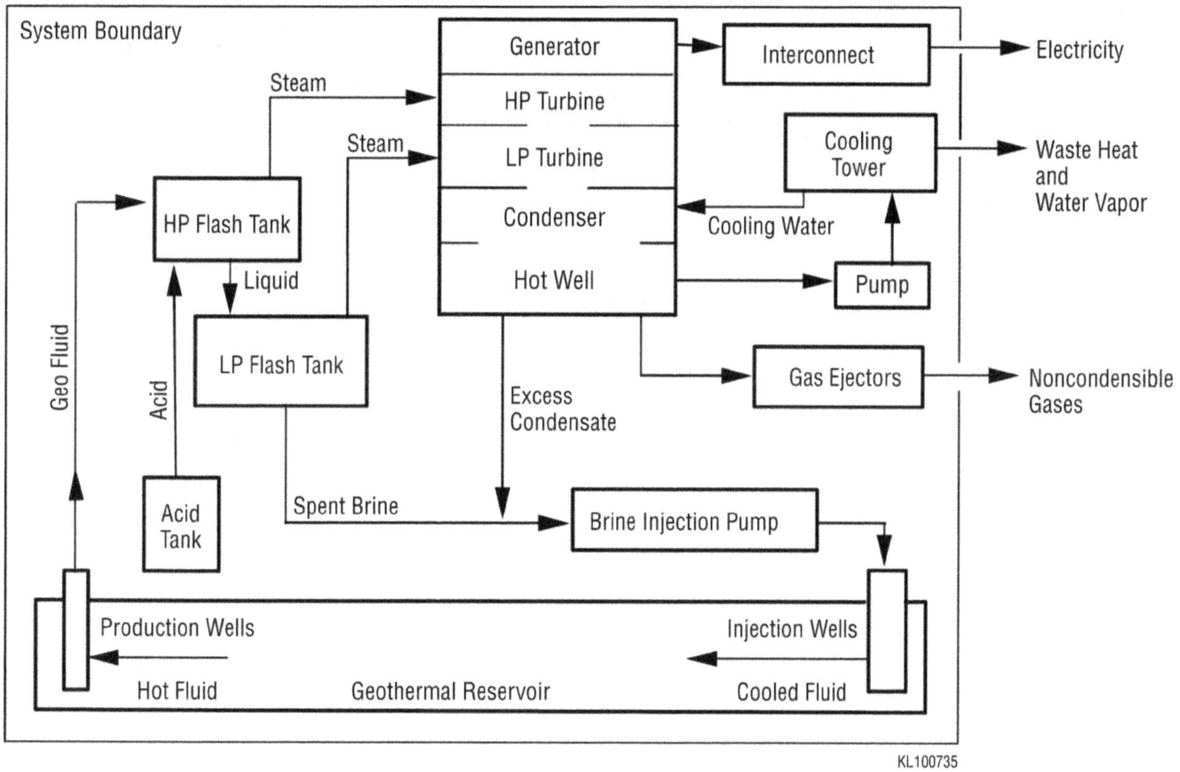

Figure D.10-13. Geothermal Hydrothermal Flashed Steam Power Plant Schematic (EERE 1997)

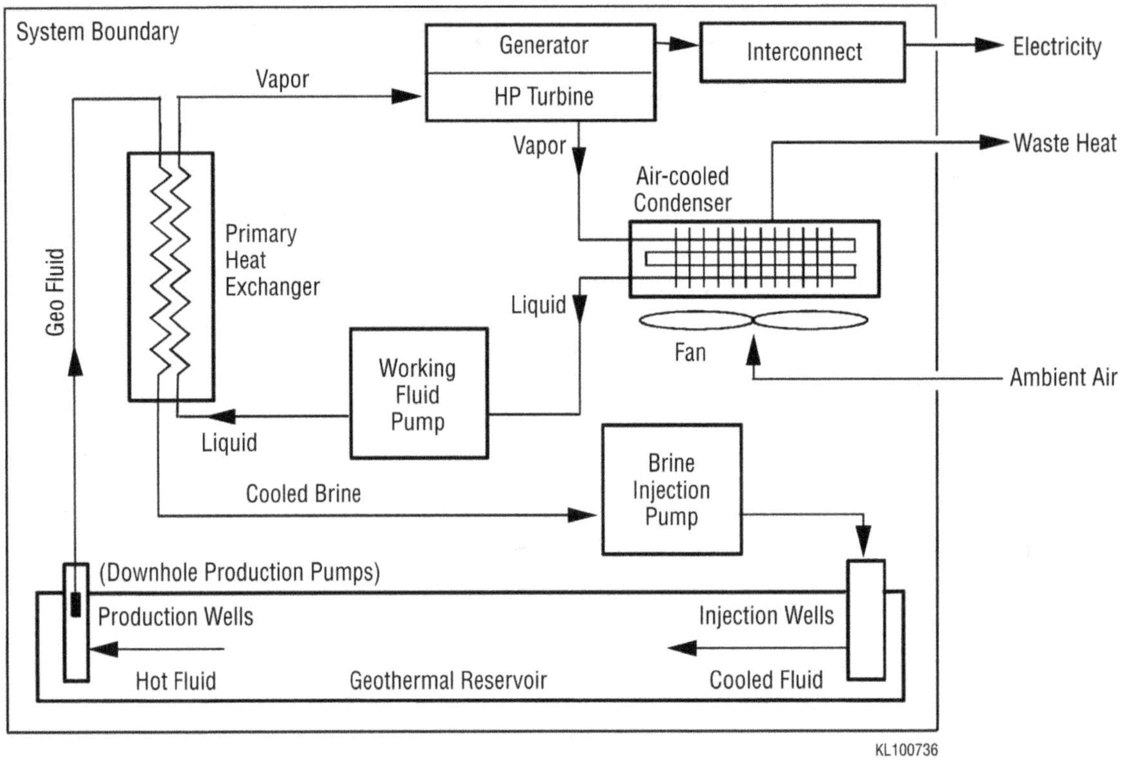

Figure D.10-14. Geothermal Hydrothermal Binary Power Plant Schematic (EERE 1997)

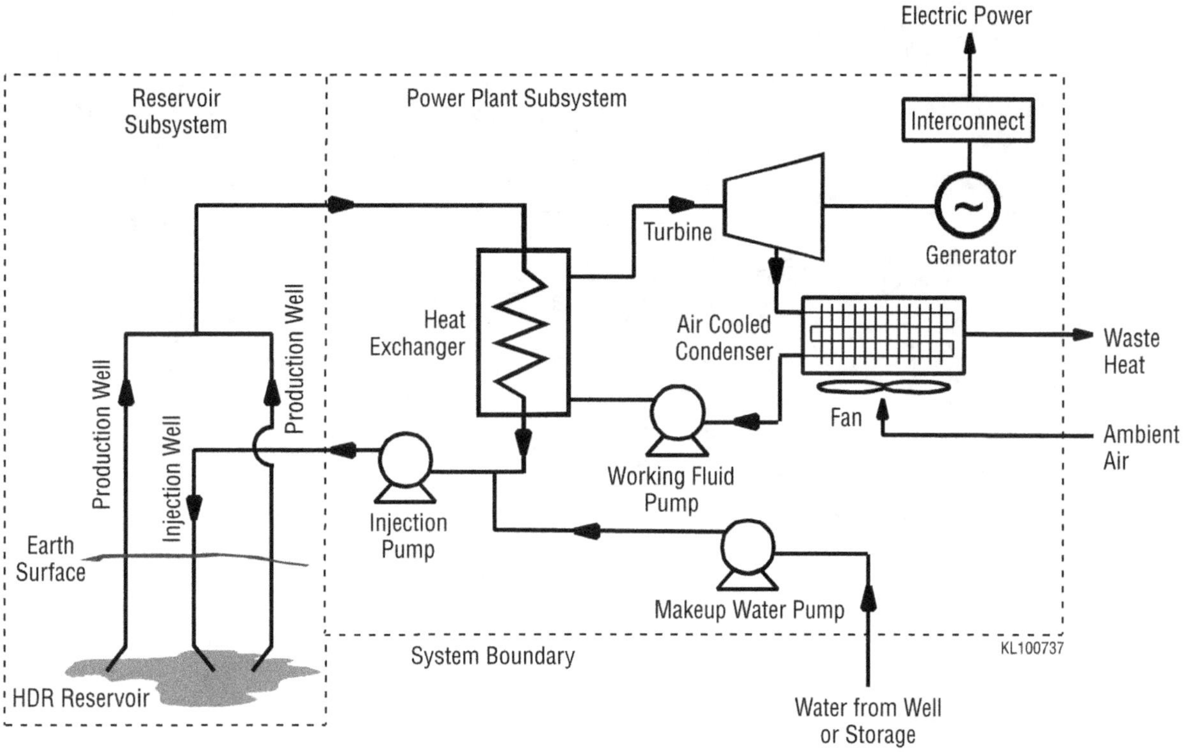

Figure D.10-15. Geothermal Hot Dry Rock Power Plant Schematic (EERE 1997)

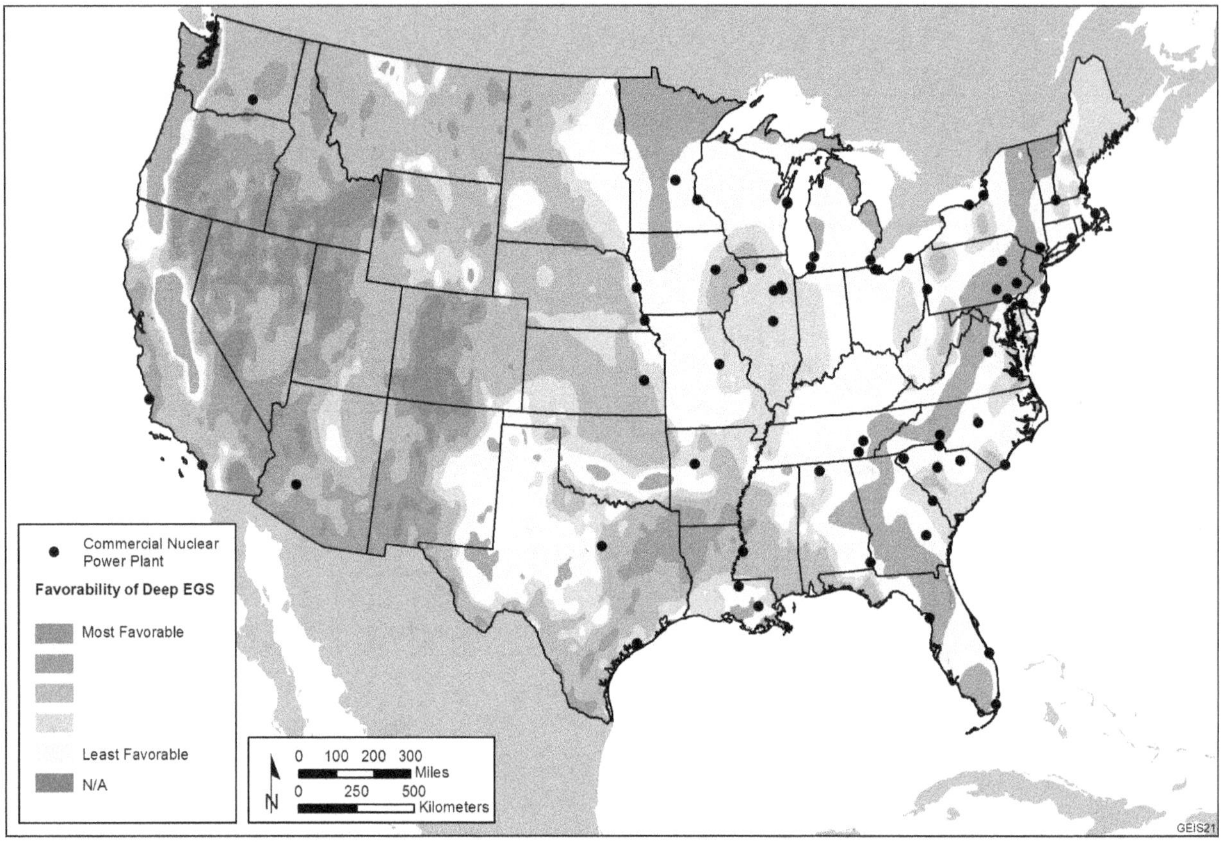

Figure D.10-16. Geothermal Resources in the 48 Contiguous United States (Adapted from NREL 2011)

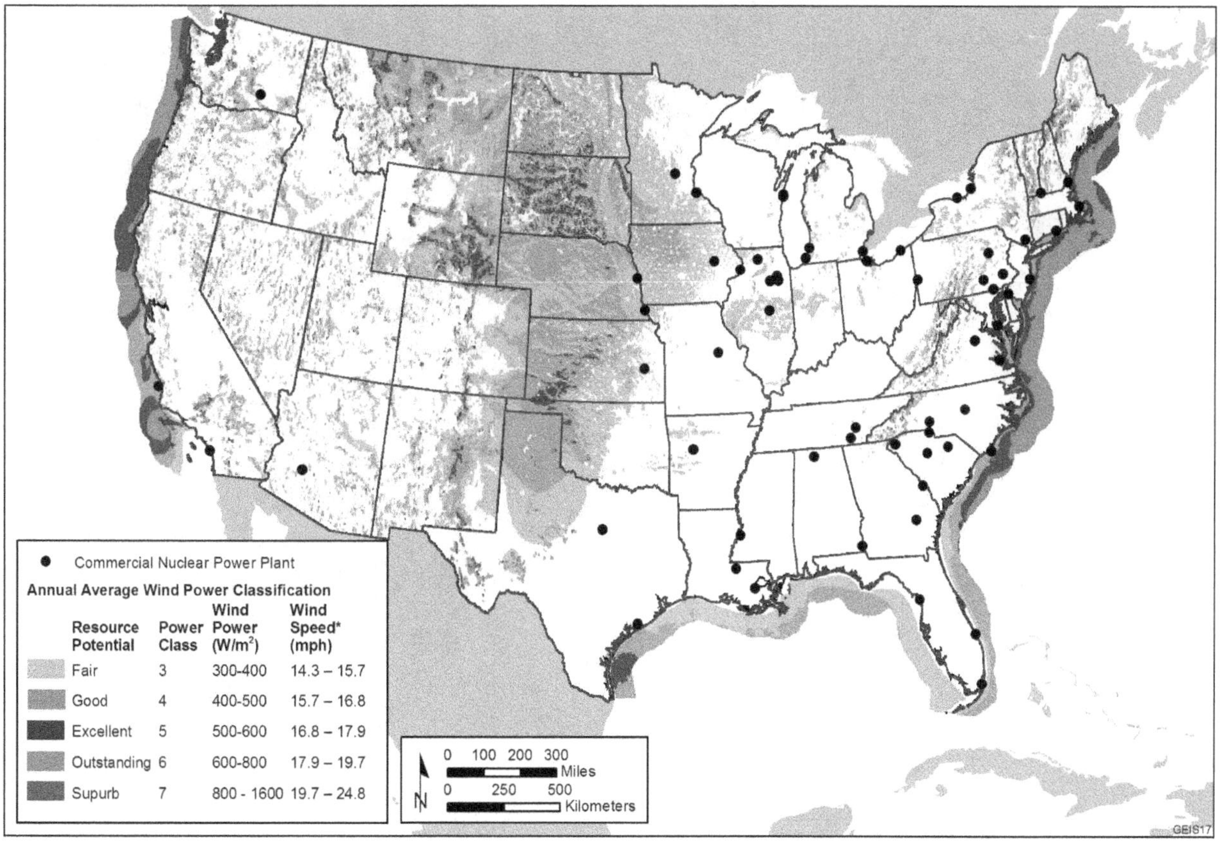

Figure D.10-17. Wind Resources in Onshore and Offshore Areas of the 48 Contiguous United States (Adapted from NREL 2011)

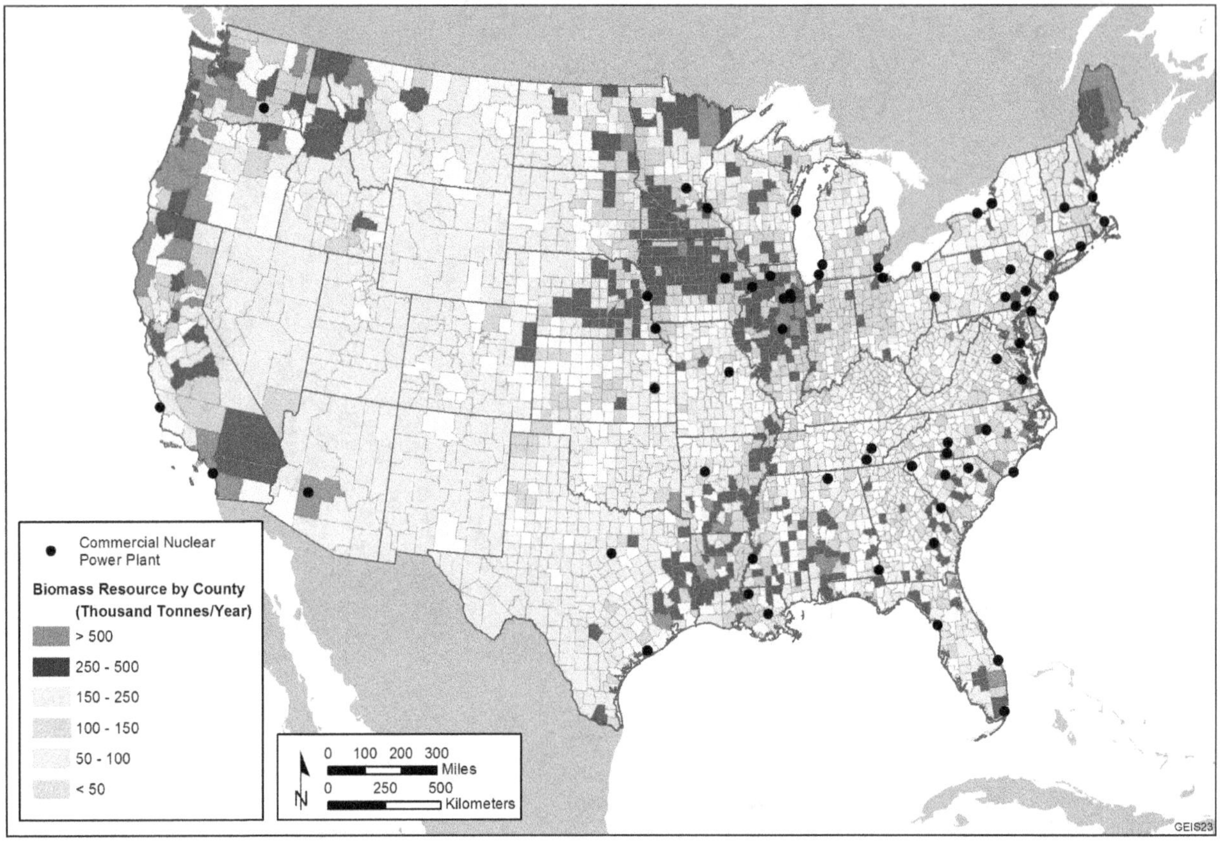

Figure D.10-18. Biomass Resources in the 48 Contiguous United States (Adapted from NREL 2011)

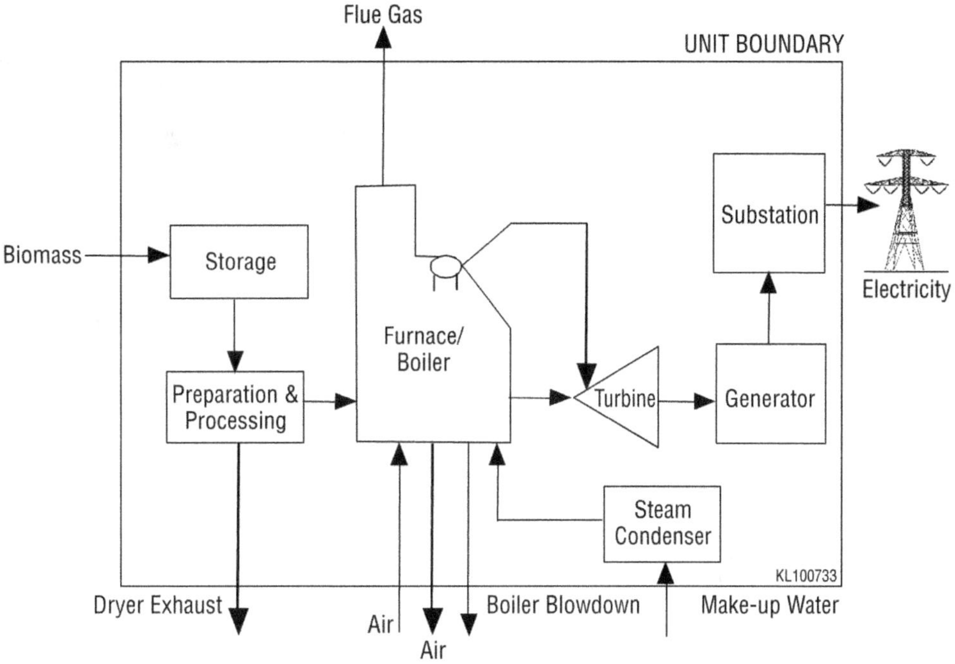

Figure D.10-19. Direct-Fire Biomass Power Plant Schematic (EERE 1997)

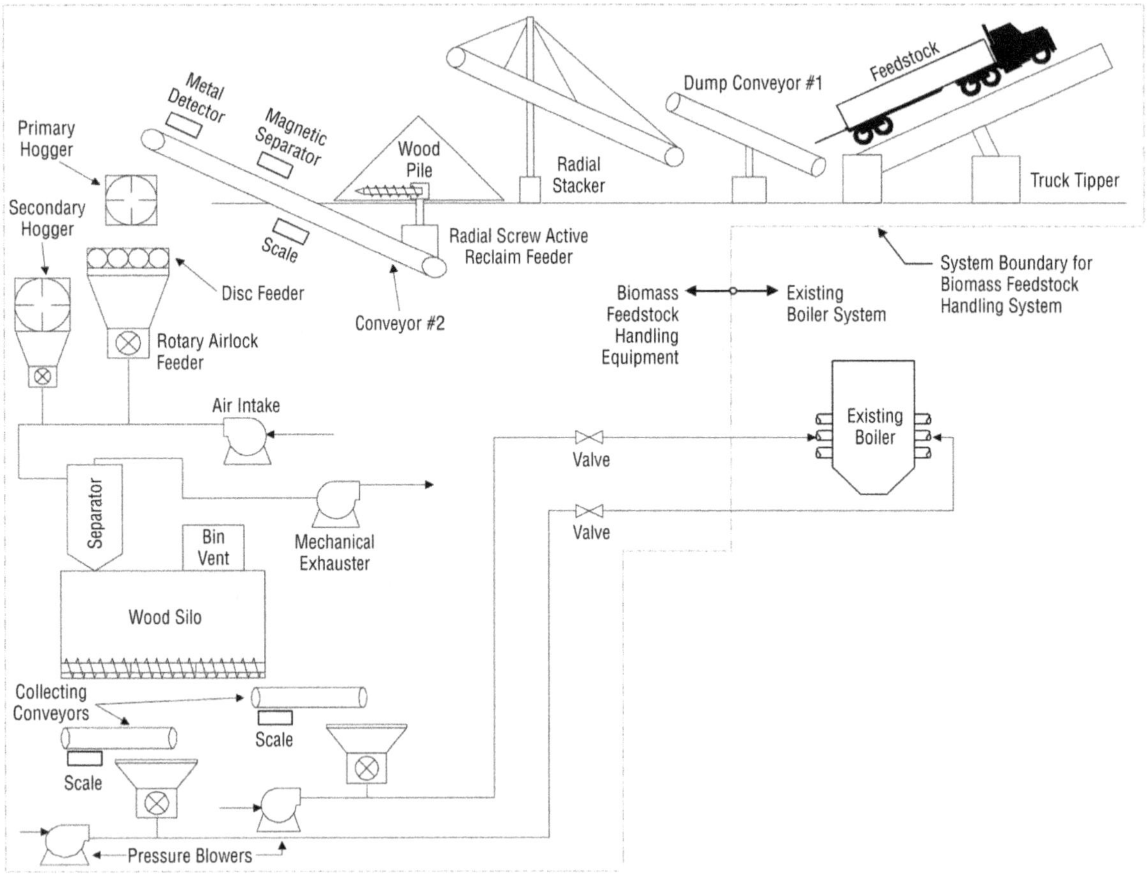

Figure D.10-20. Biomass-Coal Co-Fire Power Plant Schematic (EERE 1997)

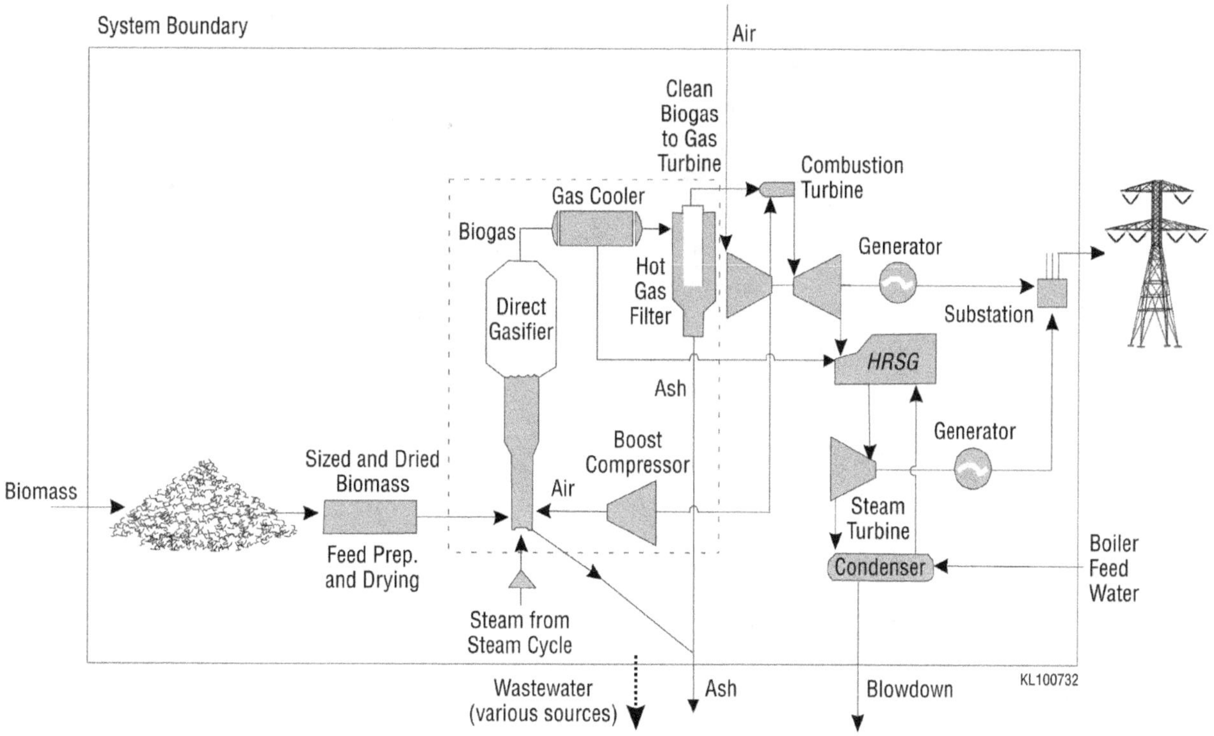

Figure D.10-21. Biomass Gasification Power Plant Schematic (EERE 1997)

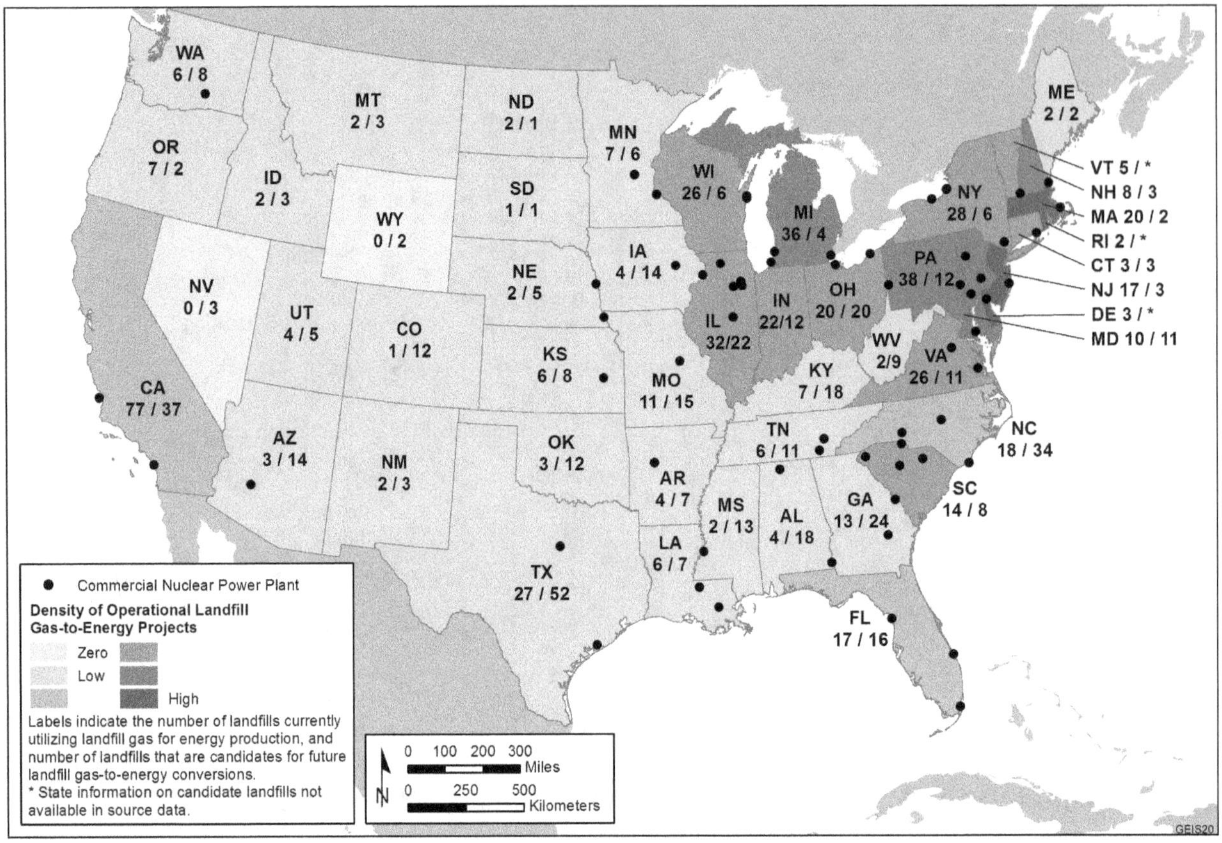

Figure D.10-22. Landfills Currently Enrolled in and Candidate Landfills for Landfill Gas-to-Energy Programs (Adapted from EPA 2011b)

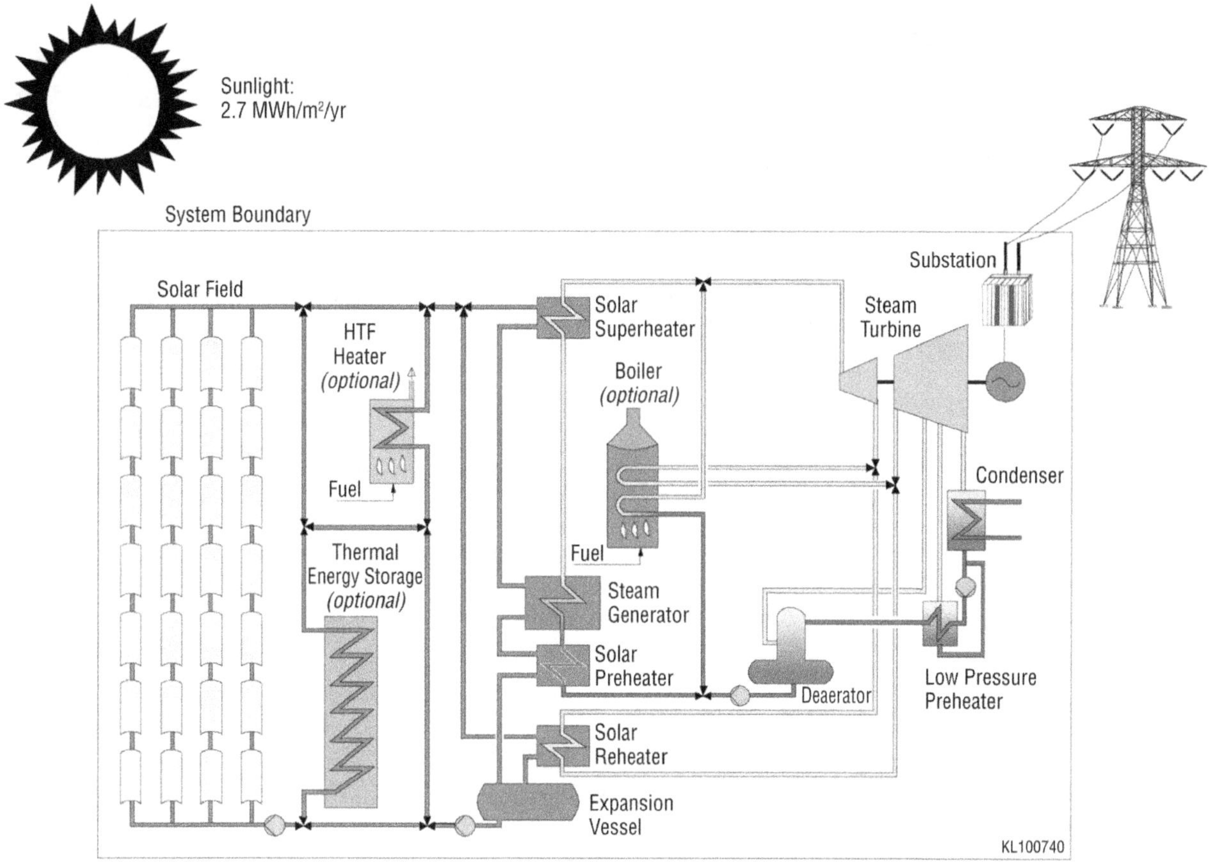

Figure D.10-23. Solar Thermal Power Trough Power Plant Schematic (EERE 1997)

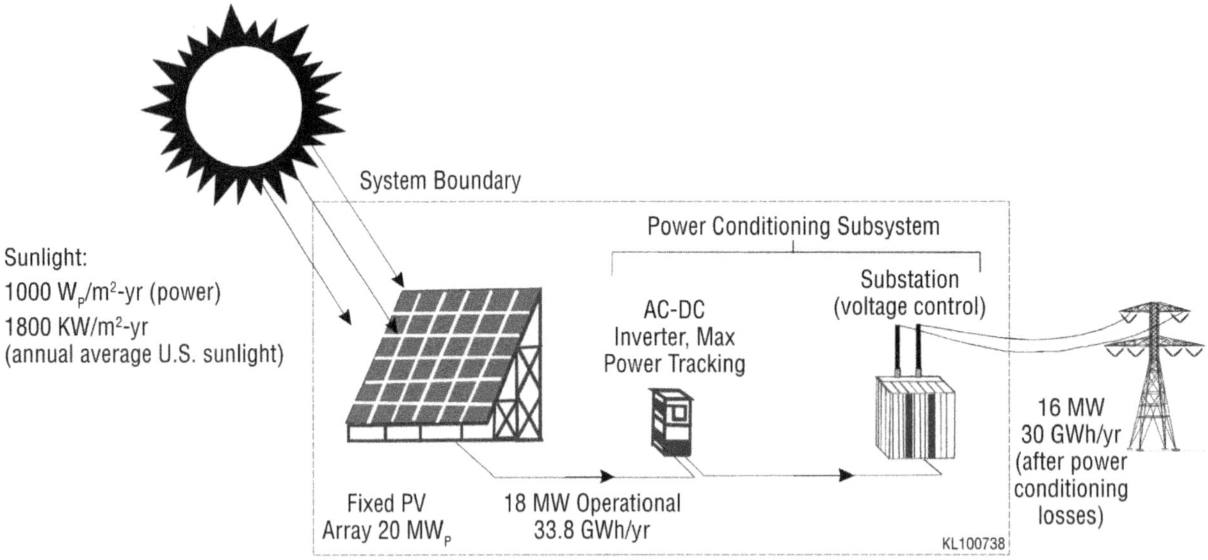

Figure D.10-24. Solar Photovoltaic Fixed Flat Plate Power Plant Schematic (EERE 1997)

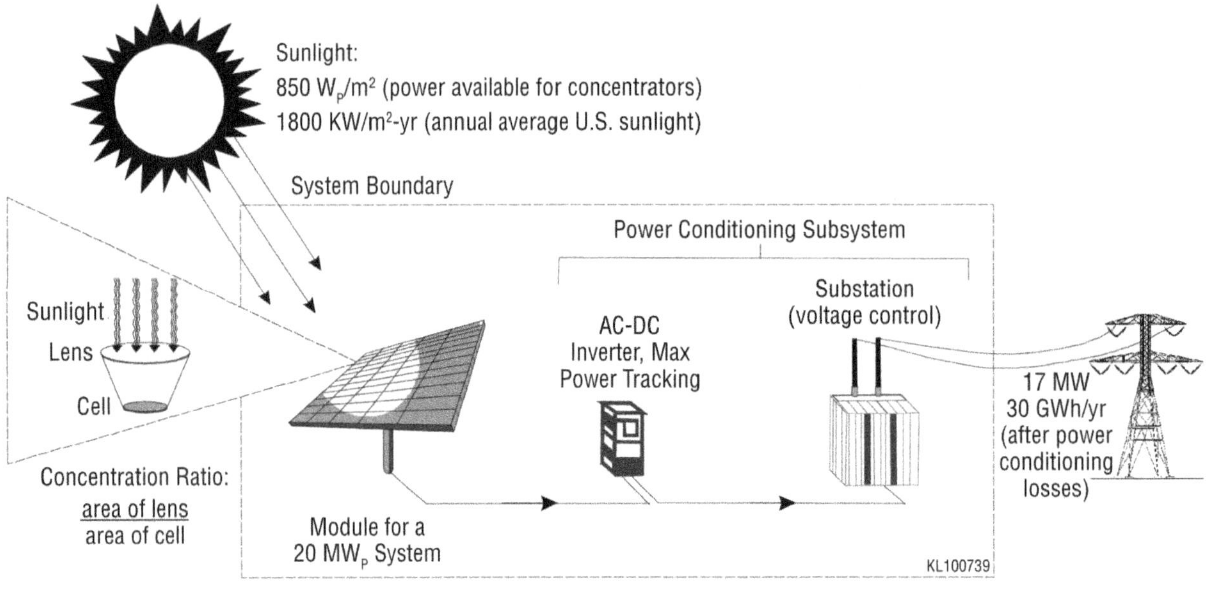

Figure D.10-25. Solar Photovoltaic Flat Plate with Concentrating Mirror Power Plant Schematic (EERE 1997)

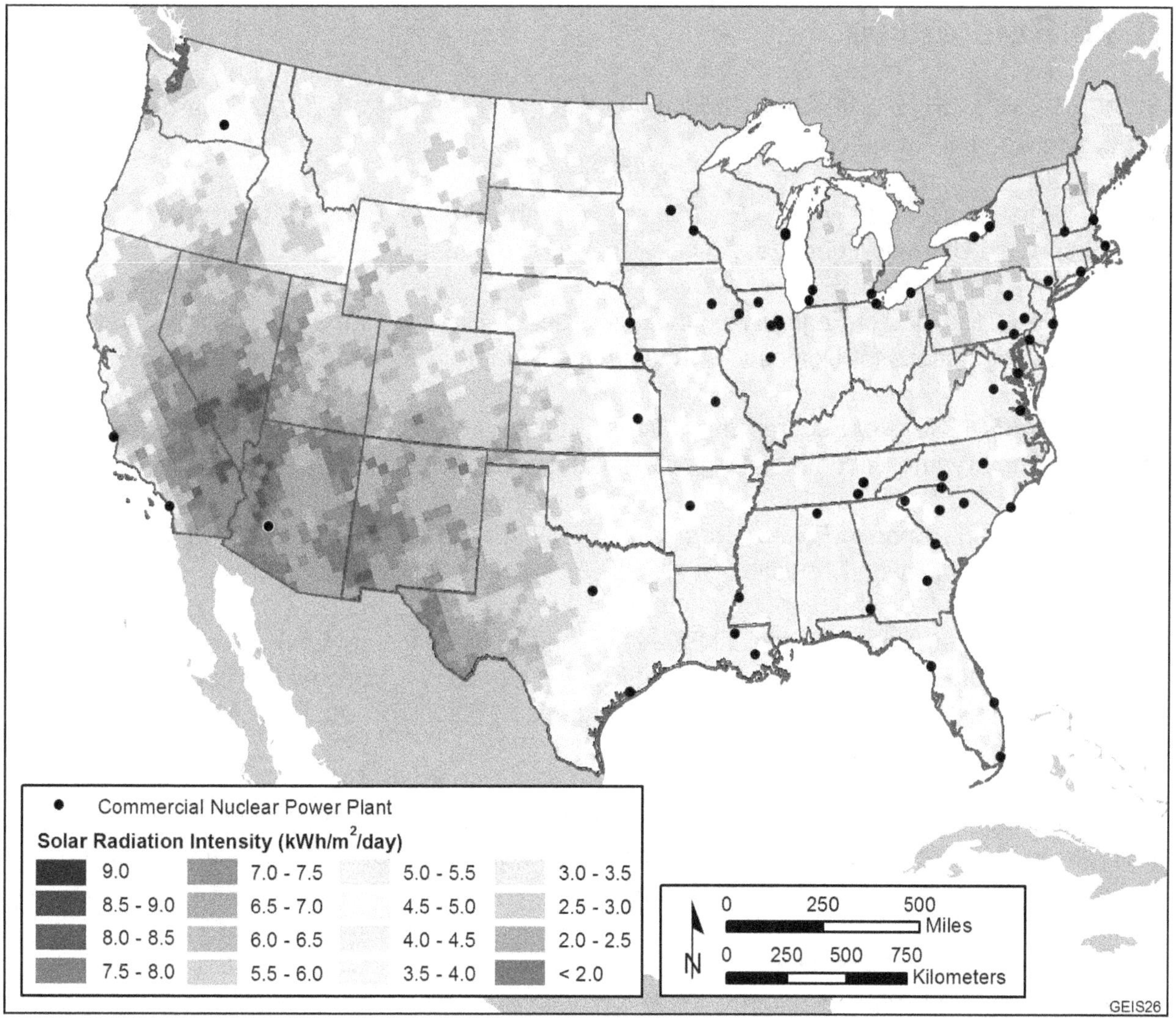

Figure D.10-26. Solar Radiation Intensity in the 48 Contiguous United States (direct normal solar radiation with two-axis tracking concentrator) (Adapted from NREL 2011)

D.11 References

10 CFR Part 20. *Code of Federal Regulations*, Title 10, *Energy*, Part 20, "Standards for Protection Against Radiation."

10 CFR Part 50. *Code of Federal Regulations*, Title 10, *Energy*, Part 50, "Domestic Licensing of Production and Utilization Facilities."

10 CFR Part 51. *Code of Federal Regulations*, Title 10, *Energy*, Part 51, "Environmental Protection Regulations for Domestic Licensing and Related Regulatory Functions."

40 CFR Part 50. *Code of Federal Regulations*, Title 40, *Protection of Environment*, Part 50, "National Primary and Secondary Ambient Air Quality Standards."

40 CFR Part 190. *Code of Federal Regulations*, Title 40, *Protection of Environment*, Part 190, "Environmental Radiation Protection Standards for Nuclear Power Operations."

Advisory Committee on the Biological Effects of Ionizing Radiation (BEIR). 1990. *BEIR-V, Health Effects of Exposure to Low Levels of Ionizing Radiation.* National Research Council, Advisory Committee on the Biological Effects of Ionizing Radiation, National Academy of Sciences, Washington, D.C.

Advisory Committee on the Biological Effects of Ionizing Radiation (BEIR). 2006. *Health Risk from Exposure to Low Levels of Ionizing Radiation.* BEIR-VII. National Research Council, National Academy of Sciences, Washington, D.C.

Burrows, S., and D.A. Hagemeyer. 2006. *Occupational Radiation Exposure at Commercial Nuclear Power Reactors and Other Facilities 2005.* NUREG-0713, U.S. Nuclear Regulatory Commission. December.

Commission for Environmental Cooperation (CEC). 1997. *Ecological Regions of North America, Toward a Common Perspective.*

Commission for Environmental Cooperation (CEC). 2006. *Ecoregions of North America, Level 1.* U.S. Environmental Protection Agency, Western Ecology Division. Available URL: http://www.epa.gov/wed/pages/ecoregions/na_eco.htm#Level%20I (Accessed November 6, 2007).

Council on Environmental Quality (CEQ). 1997. *Environmental Justice: Guidance Under the National Environmental Policy Act.* Executive Office of the President, Washington, D.C.

Cowardin, L., V. Carter, F. Golet, and E. LaRoe. 1979. *Classification of Wetlands and Deepwater Habitats of the United States, Office of Biological Services.* FWS/OBS-79/31. U.S. Fish and Wildlife Service, U.S. Department of the Interior.

Energy Efficiency and Renewable Energy (EERE). 1997. *Renewable Energy Technology Characterizations.* Topical Report TR-109496. U.S. Department of Energy, Office of Utility Technologies, Washington, D.C. December. Available URL: http://www1.eere.energy.gov/ba/pba/tech_characterizations.html (Accessed November 1, 2007).

Executive Order 12898. 1994. "Federal Actions to Address Environmental Justice in Minority and Low-Income Populations." *Federal Register* 59(32). February 16.

International Commission on Radiological Protection (ICRP). 1977. *Recommendations of the International Commission on Radiological Protection.* Publication 26. Pergamon Press, Oxford, U.K.

International Commission on Radiological Protection (ICRP). 1991. *1990 Recommendations of the International Commission on Radiological Protection.* ICRP Publication 60. Pergamon Press, Oxford, England.

National Council on Radiation Protection and Measurement (NCRP). 1993. *Risk Estimates for Radiation Protection.* Report No. 115. Bethesda, Maryland. December 31.

National Energy Technology (NETL). 2007. *Cost and Performance Baseline for Fossil Energy Plants.* DOE/NETL2007/1281, Vol. 1, Final Report, Pittsburgh, Pennsylvania.

National Renewable Energy Laboratory (NREL). 2011. *Dynamic Maps, GIS Data, and Analysis Tools—Maps.* Available URL: http://www.nrel.gov/renewable_resources, last updated May 20, 2011 (Accessed February 16, 2012).

National Renewable Energy Laboratory (NREL). 2007. *Renewable Resources Maps & Data.* Available URL: http://www.nrel.gov/renewable_resources (Accessed November 8, 2007).

Nuclear Energy Institute (NEI). 2003. *Economic Benefits of Millstone Power Station: An Economic Impact Study by the Nuclear Energy Institute.* July.

Nuclear Energy Institute (NEI). 2004a. *Economic Benefits of Diablo Canyon Power Plant: An Economic Impact Study by the Nuclear Energy Institute in Cooperation with Pacific Gas and Electric Company.* February.

Nuclear Energy Institute (NEI). 2004b. *Economic Benefits of Indian Point Energy Center: An Economic Impact Study by the Nuclear Energy Institute.* April.

Nuclear Energy Institute (NEI). 2004c. *Economic Benefits of the Duke Power-Operated Nuclear Power Plants: An Economic Impact Study by the Nuclear Energy Institute*. December.

Nuclear Energy Institute (NEI). 2004d. *Economic Benefits of Palo Verde Nuclear Generating Station: An Economic Impact Study by the Nuclear Energy Institute*. November.

Nuclear Energy Institute (NEI). 2005a. *Economic Benefits of Three Mile Island Unit 1: An Economic Impact Study by the Nuclear Energy Institute*. November.

Nuclear Energy Institute (NEI). 2005b. *Economic Benefits of Wolf Creek Generating Station: An Economic Impact Study by the Nuclear Energy Institute*. July.

Nuclear Energy Institute (NEI). 2006a. *Economic Benefits of Grand Gulf Nuclear Station : An Economic Impact Study by the Nuclear Energy Institute in Cooperation with Entergy*. December.

Nuclear Energy Institute (NEI). 2006b. *Economic Benefits of the Exelon Pennsylvania Nuclear Fleet: An Economic Impact Study by the Nuclear Energy Institute*. August.

Nuclear Energy Institute (NEI). 2006c. *Economic Benefits of PPL Susquehanna Nuclear Power Plant: An Economic Impact Study by the Nuclear Energy Institute in Cooperation with PPL Corporation*. November.

Nuclear Energy Institute (NEI). 2007. *Industry Ground Water Protection Initiative—Final Guidance Document: NEI 07-07 [Final]*. Washington, D.C. Available URL: http://adams websearch2.nrc.gov/idmws/doccontent.dll?library=PU_ADAMS^PBNTAD01&ID=072670589 (Accessed November 5, 2007).

United Nations Scientific Committee on the Effects of Atomic Radiation (UNSCEAR). 2000. *Sources and Effects of Ionizing Radiation: UNSCEAR 2000 Report to the General Assembly, with Scientific Annexes. Vol. II: Effects*. New York, New York. Available URL: http://www. unscear.org/unscear/en/publications/2000_1.html (Accessed January 31, 2008).

U.S. Department of Energy (DOE). 2002. *A Graded Approach for Evaluating Radiation Doses to Aquatic and Terrestrial Biota*. DOE Voluntary Consensus Technical Standard DOE-STD-1153-2002. Washington, D.C.

U.S. Department of Energy (DOE). 2004. *RESRAD-BIOTA: A Tool for Implementing a Graded Approach to Biota Dose Evaluation, User's Guide, Version 1*. DOE/EH-0676, ISCORS Technical Report 2004-02. Interagency Steering Committee on Radiation Standards, Office of Air, Water and Radiation Protection Policy and Guidance, Washington, D.C. January. Available URL: http://web.ead.anl.gov/resrad/documents/RESRAD-BIOTA_Manual_Version_1.pdf (Accessed November 26, 2007).

U.S. Environmental Protection Agency (EPA). 1999. *Cancer Risk Coefficients for Environmental Exposure to Radionuclides*. Federal Guidance Report No. 13, EPA 402-R-99-001. Office of Radiation and Indoor Air, Washington, D.C. September.

U.S. Environmental Protection Agency (EPA). 2007. *Level III Ecoregions of the Conterminous United States*. National Health and Environmental Effects Research Laboratory. Available URL: http://www.epa.gov/wed/pages/ecoregions/level_iii.htm (Accessed November 6, 2007).

U.S. Environmental Protection Agency (EPA). 2011a. *National Ambient Air Quality Standards (NAAQS)*. Available URL: http://www.epa.gov/air/criteria.html, last updated November 8, 2011 (Accessed February 3, 2012).

U.S. Environmental Protection Agency (EPA). 2011b. *Energy Projects and Candidate Landfills*. Available URL: http://www.epa.gov/lmop/projects-candidates/index.html#map-area, last updated July 25, 2011 (Accessed February 17, 2012).

U.S. Fish and Wildlife Service (USFWS). 2007. *National Wetlands Inventory*. U.S. Department of the Interior. Available URL: http://wetlandsfws.er.usgs.gov/NWI/index.html (Accessed November 7, 2007).

U.S. Geological Survey (USGS). 2003. *National Land Cover Database*. Sioux Falls, South Dakota.

U.S. Nuclear Regulatory Commission (NRC). 1977a. *Estimating Aquatic Dispersion of Effluents from Accidental and Routine Reactor Releases for the Purpose of Implementing Appendix I*. Regulatory Guide 1.113, Revision 1. Office of Standards Development. April.

U.S. Nuclear Regulatory Commission (NRC). 1977b. *Method of Estimating Atmospheric Transport and Dispersion of Gaseous Effluents in Routine Releases from Light-Water-Cooled Reactors*. Regulatory Guide 1.111, Revision 1. Office of Standards Development. July.

U.S. Nuclear Regulatory Commission (NRC). 1977c. *Calculation of Annual Doses to Man from Routine Releases of Reactor Effluents for the Purpose of Evaluating Compliance with 10 CFR Part 50, Appendix I.* Regulatory Guide 1.109, Revision 1. Office of Standards Development. October.

U.S. Nuclear Regulatory Commission (NRC). 1996. *Generic Environmental Impact Statement for License Renewal of Nuclear Plants*. NUREG-1437, Vols. 1 and 2, Washington, D.C.

U.S. Nuclear Regulatory Commission (NRC). 2002. *Final Generic Environmental Impact Statement on Decommissioning Nuclear Facilities, Supplement 1, Regarding the Decommissioning of Nuclear Power Reactors*. NUREG-0586, Vols. 1 and 2, Washington, D.C.

U.S. Nuclear Regulatory Commission (NRC). 2005. *Staff Review of the National Academies Study of the Health Risks from Exposure to Low Levels of Ionizing Radiation (BEIR VII).* Commission Paper SECY-05-0202. October 29. ADAMS Accession No. ML052640532.

U.S. Nuclear Regulatory Commission (NRC). 2006. *Liquid Radioactive Release Lessons Learned Task Force Final Report*. Available URL: http://adamswebsearch2.nrc.gov/idmws/doccontent.dll?library=PU_ADAMS^PBNTAD01&ID=062770207 (Accessed October 17, 2007).

Appendix E

Environmental Impact of Postulated Accidents

Appendix E

Environmental Impact of Postulated Accidents

E.1 Introduction

Chapter 5 of the *Generic Environmental Impact Statement for License Renewal of Nuclear Plants (*GEIS), Volumes 1 and 2 (NRC 1996, 1999)[(a)] assessed the impacts of postulated accidents at nuclear power plants on the environment. The postulated accidents included design-basis accidents and severe accidents (e.g., those with core damage). The impacts considered included:

- Dose and health effects of accidents (Sections 5.3.3.2 through 5.3.3.4);
- Economic impacts of accidents (Section 5.3.3.5); and
- Effect of uncertainties on the results (Section 5.3.4).

The estimated impacts were based on the analysis of severe accidents at 28 nuclear power plant sites[(b)] as reported in the environmental impact statements (EISs) and/or final environmental statements (FESs) prepared for each of the 28 plants in support of their operating licenses. With few exceptions, the severe accident analyses were limited to consideration of reactor accidents caused by internal events. The 1996 GEIS addressed the impacts from external events qualitatively.[(c)] The severe accident analysis for the 28 sites was extended to the remainder of plants whose EISs did not consider severe accidents (since such analyses were not required at the time the other plants' EISs were prepared). The estimates of environmental impact contained in the 1996 GEIS used 95th percentile upper confidence bound (UCB) estimates whenever available. This approach provides conservatism to cover

---

(a) The GEIS was originally issued in 1996. Addendum 1 to the GEIS was issued in 1999. Hereafter, all references to the "1996 GEIS" include the original GEIS and Addendum 1.

(b) The 28 sites are listed in Table 5.1 of the 1996 GEIS. There are a total of 44 units included in this list (at the 28 sites), but 4 of these units never operated (Grand Gulf 2, Harris 2, Perry 2, and Seabrook 2). For the purpose of this appendix, this list will be referred to as containing 28 nuclear power plants, but when mean values are calculated for this subset of nuclear power plants, all 40 units that operated are considered.

(c) See Section 5.3.3.1 of the 1996 GEIS, including a brief discussion of the external event risk assessments conducted by the staff prior to 1996, which included assessments for Zion 1 & 2, Indian Point 2 & 3, Limerick 1 & 2, Surry 1, Peach Bottom 2, and Millstone 3.

uncertainties, as described in Section 5.3.3.2.2 of the 1996 GEIS. The 1996 GEIS concluded that the probabilistically weighted impacts were small compared to other risks to which the populations surrounding nuclear power plants are routinely exposed.

The focus of this revision is on severe accidents since the impacts from design-basis accidents are SMALL and, as stated in Section E.3 of this revision, the U.S. Nuclear Regulatory Commission's (NRC's) assessment remains unchanged. Since the NRC's understanding of severe accident risk has evolved since issuance of the 1996 GEIS, this appendix assesses more recent information on severe accidents that might alter the conclusions in Chapter 5 of the 1996 GEIS. This revision considers how these developments would affect the conclusions in the 1996 GEIS and provides comparative data where appropriate. This revision does not attempt to provide new quantitative estimates of severe accident impacts. In addition, the revision only covers one initial license renewal period for each plant (as did the 1996 GEIS). Thus, the population projections, meteorology, and exposure indices used in the 1996 GEIS are assumed to remain unchanged for purposes of this analysis.

Finally, the format of this appendix follows the same format as used in Chapter 5 of the 1996 GEIS, including a discussion on uncertainties and severe accident mitigation alternatives (SAMAs).

E.2 Plant Accidents

A general description of plant accidents is contained in Section 5.2 of the 1996 GEIS. This description covered:

- The general characteristics of accidents;
- Fission product characteristics;
- Meteorological considerations;
- Exposure pathways;
- Adverse health effects;
- Avoiding adverse health effects;
- Accident experience and observed impacts;

- Mitigation of accident consequences; and

- Emergency preparedness.

This description is still valid and thus remains unchanged. Section 5.2 of the 1996 GEIS also mentions that as of 1990, there have been approximately 1,300 reactor-years of experience to support the safety of U.S. nuclear power plants. As with any technology, experience generally leads to improved plant performance and public safety. As of 2011, there has been approximately an additional 2,000 reactor-years of experience in the United States. This additional experience has contributed to improved plant performance (e.g., as measured by trends in plant-specific performance indicators), a reduction in operating events, and lessons learned that improve the safety of all of the operating nuclear power plants. Other examples of items contributing to improved safety include:

- Implementation of plant improvements identified through the Individual Plant Examination (IPE) and Individual Plant Examination: External Events (IPEEE) programs (e.g., strengthening of seismic supports; enhanced fire brigade training) (NRC 2003);

- Identification of specific aging mechanisms (e.g., cables; irradiation-assisted stress corrosion cracking) and development of programs to monitor and control these mechanisms (NRC 2001c);

- NRC staff actions on generic safety issues (e.g., Generic Safety Issue 191 on sump performance) (NRC 2008e); and

- Implementation of the NRC's Interim Compensatory Measures (ICMs) Order following the September 2001 terrorist attacks.[d]

Thus, the performance and safety record of nuclear power plants operating in the United States continues to improve. This is also confirmed by analysis which indicates that, in many cases, improved plant performance and design features have resulted in reductions in initiating event frequency, core damage frequency, and containment failure frequency.[e]

(d) The safety evaluations (SEs) for the operating license amendments associated with implementation of Section B.5.b. of Commission Order EA-02-026 provide background related to the implementation of particular portions of the ICMs. As an example, the reader is referred to the SE associated with Brunswick Steam Electric Plant, Units 1 and 2 (NRC 2007a).

(e) This statement is based on industry performance data provided in the NRC's *2007-2008 Information Digest* (NRC 2007b) and on the NRC's website (NRC 2008c), as well as information contained in Chapter 5 of site-specific EISs (the NUREG-1437 series of supplements).

E.2.1 Fukushima Earthquake and Tsunami

On March 11, 2011, a massive earthquake off the east coast of Honshu, Japan, produced a devastating tsunami that struck the coastal town of Fukushima. The six-unit Fukushima Dai-ichi nuclear power plant was directly impacted by these events. The resulting damage caused the failure of several of the units' safety systems needed to maintain cooling water flow to the reactors. As a result of the loss of cooling, the fuel overheated, and there was a partial meltdown of the fuel contained in several of the reactors. Damage to the systems and structures containing reactor fuel resulted in the release of radioactive material to the surrounding environment.

In response to the earthquake, tsunami, and resulting reactor accidents at Fukushima Dai-ichi (hereafter referred to as the "Fukushima events"), the Commission directed the staff to convene an agency task force of senior leaders and experts to conduct a methodical and systematic review of the relevant NRC regulatory requirements, programs, and processes, including their implementation, and to recommend whether the agency should make near-term improvements to its regulatory system. As part of the short-term review, the task force concluded that, while improvements are expected to be made as a result of the lessons learned from the Fukushima events, the continued operation of nuclear power plants and licensing activities for new plants do not pose an imminent risk to public health and safety (NRC 2011).

During the time that the task force was conducting its review, groups of individuals and non-governmental organizations petitioned the Commission to suspend all licensing decisions in order to conduct a separate, generic National Environmental Policy Act (NEPA) analysis to determine whether the Fukushima events constituted "new and significant information" under NEPA that must be analyzed as part of environmental reviews. The Commission found the request premature and noted, "In short, we do not know today the full implications of the [Fukushima] events for U.S. facilities."[f] However, the Commission found that if "new and significant information comes to light that requires consideration as part of the ongoing preparation of application-specific NEPA documents, the agency will assess the significance of that information, as appropriate."[g] The Federal courts of appeal and the Commission have interpreted NEPA such that an EIS must be updated to include new information only when that

---

(f) *Union Electric Co. d/b/a Ameren Missouri* (Callaway Plant, Unit 2), CLI-11-05, 74 NRC141, 167 (Sept. 9, 2011).

(g) *Id.*

new information provides "a seriously different picture of the environmental impact of the proposed project from what was previously envisioned."[(h)]

In the context of the GEIS, the Fukushima events are considered a severe accident (i.e., a type of accident that may challenge a plant's safety systems at a level much higher than expected) and more specifically, a severe accident initiated by an event external to the plant. The 1996 GEIS concluded that risks from severe accidents initiated by external events (such as an earthquake) could have potentially high consequences but found that external events are adequately addressed through a consideration of a severe accident initiated by an internal event (such as a loss of cooling water). Therefore, an applicant for license renewal need only analyze the environmental impacts from an internal event in order to adequately characterize the environmental impacts from either type of event. Prior to the Fukushima events, this GEIS examined more recent and up-to-date information regarding external events and concluded that the analysis in the 1996 GEIS remains valid.

As of the publication date of this GEIS, the NRC's evaluation of the consequences of the Fukushima events is ongoing. As such, the NRC will continue to evaluate the need to make improvements to existing regulatory requirements based on the task force report and additional studies and analyses of the Fukushima events as more information is learned. To the extent that any revisions are made to NRC regulatory requirements, they would be made applicable to nuclear power reactors regardless of whether or not they have a renewed license. Therefore, no additional analyses have been performed in this GEIS as a result of the Fukushima events. In the event that the NRC identifies information from the Fukushima events that constitutes new and significant information with respect to the environmental impacts of license renewal, the NRC will discuss that information in its site-specific supplemental EISs (SEISs) to the GEIS, as it does with all such new and significant information.

E.3 Accident Risk and Impact Assessment

The environmental impacts from design-basis accidents and severe accidents are assessed in Sections 5.3.2 and 5.3.3 of the 1996 GEIS, respectively. As stated in Section 5.3.2, the environmental impact from design-basis accidents was assessed in the individual plant-specific EISs at the time of the initial license application review. Since the licensee is required to

---

(h) *Id.* at 167-68 *quoting Hydro Resources, Inc.* (2929 Coors Road, Suite 101, Albuquerque, NM 87120), CLI-99-22, 50 NRC 3, 14 (1999) (citing *Marsh v. Oregon Natural Resources Council*, 490 U.S. 360, 373 (1989)). The Commission also noted that it can modify a facility's operating license outside of a renewal proceeding and made clear that "it will use the information from these activities to impose any requirement it deems necessary, irrespective of whether a plant is applying for or has been granted a renewed operating license." *Id.* at 164 *quoting Pilgrim & Indian Point:* Entergy's Answer Opposing Petition to Suspend Pending Licensing Proceedings (May 2, 2011) at 3.

maintain the plant within acceptable design and performance criteria, including during any license renewal term, these impacts are not expected to change. Therefore, additional assessment of the environmental impacts from design-basis accidents is not necessary, and the bulk of the 1996 GEIS evaluation focused on the environmental impact of severe accidents.

To assess the impacts from the airborne pathway, the 1996 GEIS relied on severe accident analyses provided in the EISs for the more recent sites. Table 5-1 in the 1996 GEIS lists the 28 nuclear power plants that included severe accident analyses in their plant-specific EISs. These plant-specific EISs used site-specific meteorology, land topography, population distributions, and offsite emergency response parameters, along with generic or plant-specific source terms, to calculate offsite health and economic impacts. The offsite health effects included those from airborne releases of radioactive material and contamination of surface water and groundwater.

The 1996 GEIS used the environmental impact information from the 28 plant-specific EISs and a metric called the exposure index to (1) scale up the radiological impact of severe accidents on the population due to demographic changes from the time the original EIS[(i)] was done until the year representing the mid-license renewal period and (2) estimate the severe accident environmental impacts for the earlier plants (whose EISs did not include a quantitative assessment of severe accidents). The exposure index method uses the projected population distribution around each nuclear power plant site at the middle of its license renewal period and meteorology data for each site to provide a measure of the degree to which the population would be exposed to the release of radioactive material resulting from a severe accident (i.e., the exposure index method weights the population in each of 16 sectors around a nuclear power plant by the fraction of time the wind blows in that direction on an annual basis). The exposure index metric was also used to project economic impacts at the mid-year of the license renewal period. A more detailed description of the exposure index method is contained in Appendix G of the 1996 GEIS. The use of the exposure index method remains valid.

Since 1996, developments in plant operation and accident analysis have taken place that could affect the assumptions made in the 1996 GEIS. These changes are grouped into the following areas and are each covered in the indicated section of this revision:

- Internal event risk (Section E.3.1);
- External event risk (Section E.3.2);

(i) The term "original EIS" describes an EIS issued by the NRC that is associated with the issuance of a plant's initial operating license. This term is used in this appendix to differentiate it from an EIS prepared in conjunction with a license renewal environmental review.

- Updates in the quantification of accident source terms (Section E.3.3);
- Increases in licensed reactor power levels, i.e., power uprates (Section E.3.4);
- Increases in fuel burnup levels (Section E.3.5);
- Consideration of reactor accidents at low power and shutdown conditions (Section E.3.6);
- Consideration of accidents in spent fuel pools (Section E.3.7); and
- The BEIR VII report on the risk of fatal cancers posed by exposure to radiation (Section E.3.8).

Sections discussing uncertainties, SAMAs, and conclusions are also provided.

As discussed in the Section 5.3.3.1 of the 1996 GEIS, the environmental impacts from security-related events were not considered in that document. As stated, these types of events are addressed via deterministic criteria in Title 10, Part 73, of the *Code of Federal Regulations* (10 CFR Part 73), rather than by risk assessments. The regulatory requirements under 10 CFR Part 73 provide reasonable assurance that the risk from sabotage is small. This section goes on to state:

> Although the threat of sabotage events cannot be accurately quantified, the Commission believes that acts of sabotage are not reasonably expected. Nonetheless, if such events were to occur, the Commission would expect that resultant core damage and radiological releases would be no worse than those expected from internally initiated events.

The NRC continues to take this position. As a result of the terrorist attacks of September 11, 2001, the NRC conducted a comprehensive review of the agency's security program and made further enhancements to security at a wide range of NRC-regulated facilities. These enhancements included significant reinforcement of the defense capabilities for nuclear facilities, better control of sensitive information, enhancements in emergency preparedness to further strengthen NRC's nuclear facility security program, and implementation of mitigating strategies to deal with postulated events potentially causing loss of large areas of the plant due to explosions or fires, including those that an aircraft impact might create. These measures are outlined in greater detail in NUREG/BR-0314 (NRC 2004), NUREG-1850 (NRC 2006a), and Sandia National Laboratory's "Mitigation of Spent Fuel Loss-of-Coolant Inventory Accidents and Extension of Reference Plant Analyses to Other Spent Fuel Pools" (NRC 2006b).

The NRC routinely assesses threats and other information provided by a variety of Federal agencies and sources. The NRC also ensures that licensees meet appropriate security-level requirements. The NRC will continue to focus on prevention of terrorist acts for all nuclear facilities and will not focus on site-specific evaluations of speculative environmental impacts resulting from terrorist acts. While these are legitimate matters of concern, the NRC will continue to address them through the ongoing regulatory process as a current and generic regulatory issue that affects all nuclear facilities and many of the activities conducted at nuclear facilities. The issue of security and risk from malevolent acts at nuclear power facilities is not unique to facilities that have requested a renewal of their licenses (NRC 2006a).

Malevolent acts remain speculative and beyond the scope of a NEPA review. NEPA requires that there be a "reasonably close causal relationship" between the federal agency action and the environmental consequences. The environmental impact of a terrorist attack is too far removed from the natural, or expected, consequences of a license renewal action to warrant consideration under NEPA. However, as noted above, in the event of a terrorist attack, the consequences of such an attack would be no worse than an internally initiated severe accident, which has already been analyzed.

In a decision dated June 2, 2006, *San Luis Obispo Mothers for Peace v. NRC*, 449 F.3d 1016, 1028 (9th Cir. 2006) the U.S. Court of Appeals for the Ninth Circuit held that NRC could not categorically refuse to consider the consequences of a terrorist attack under NEPA and remanded the case to NRC. On remand, the Commission adjudicated the intervenors' claim that the NRC staff had not adequately assessed the environmental consequences of a terrorist attack on the Diablo Canyon Power Plant's proposed facility for storing spent nuclear fuel in dry casks. *See, Pacific Gas & Electric Co.,* (Diablo Canyon Power Plant Independent Spent Fuel Storage Installation), CLI-08-26, 68 NRC 509 (2009). The Commission ultimately determined that an EIS was not required in order to address land contamination and latent health effect issues (Diablo Canyon, CLI-08-26, 68 NRC at 521). Further, the Commission concluded that the staff's final, supplemental environmental assessment and finding of no significant impact, the adjudicatory record of the case, and its supervisory review of the non-public information underlying portions of the staff's analyses, satisfied the agency's NEPA obligations. *Id.* 525-26. The staff had found that even the most severe, plausible terrorist attack of those examined would not cause immediate or latent health effects. The staff also found that such an attack was improbable, but if one occurred, the likelihood of significant radioactive release was very low because the nature of the Diablo Canyon casks and site. *Id.* at 521. The U.S. Court of Appeals for the Ninth Circuit upheld the Commission's determination on appeal. *San Luis Obispo Mothers for Peace v. NRC*, 645 F.3d 1109, 1120-21 (9th Cir. 2011).

The Commission stated that it will adhere to the Ninth Circuit decision when considering licensing actions for facilities subject to the jurisdiction of that Circuit. *See Pacific Gas and Electric Co., (Diablo Canyon Power Plant Independent Spent Fuel Storage Installation*), CLI-07-11, 65 NRC 118 (2007). However, the Commission decided against applying that holding to all licensing proceedings nationwide. In one such proceeding, *Amergen Energy Co. LLC* (Oyster Creek Nuclear Generating Station), CLI-07-8, 65 NRC 124, 128-29 (2007), the New Jersey Department of Environmental Protection contended that NEPA requires an analysis of a terrorist attack. The NRC found that NEPA "imposes no legal duty on the NRC to consider intentional malevolent acts" because such acts are "too far removed from the natural or expected consequences of agency action." *Id.* at 129 (quoting the Board decision). The NRC also found that a terrorism review would be redundant because (1) "the NRC has undertaken extensive efforts to enhance security at nuclear facilities," which it characterized as the best mechanism to protect the public; *id.* at 130; (2) the GEIS had addressed the issue and concluded that "the core damage and radiological release from [terrorist] acts would be no worse than the damage and release to be expected from internally initiated events." On appeal, the Third Circuit agreed with the NRC and denied the petition. *See NJDEP v. NRC and Amergen Energy Co, LLC,* (Case No. 07-2271), 561 F.3rd 132 (3rd Cir. 2009). The Court found that, "the NRC correctly concluded that the relicensing of Oyster Creek does not have a 'reasonably close causal relationship' with the environmental effects that would be caused in the event of a terrorist attack." 561 F.3d at 143.

The Third Circuit disagreed with the Ninth Circuit's application of the relevant Supreme Court decisions. Instead, as the Commission had originally held, the Third Circuit concluded that the issuance of a facility license—here, the issuance of the 20-year extension for the Oyster Creek license—would not be the "proximate cause" of a terrorist attack on the facility.

Moreover, the Third Circuit noted that the GEIS for License Renewal had reviewed the possible impacts of a sabotage event, which is a form of terrorism. The GEIS found that the consequences of a sabotage event would be no worse than those expected from an internally initiated severe accident. The Third Circuit noted that the petitioner in the case before it (the State of New Jersey) had failed to demonstrate that the results of a terrorist attack would be any different than those of a severe accident, which had already been analyzed. The Third Circuit also noted that the NRC had prepared a site-specific EIS addressing the mitigation of severe accidents at Oyster Creek. As a result, the Third Circuit found that, even if the Commission were required to analyze the impacts of a terrorist attack, the NRC had prepared both generic and site-specific analyses of the impacts of a terrorist attack at Oyster Creek, and that the Petitioner had not shown that the NRC could evaluate the risks more meaningfully than it had already done.

Subsequent to the Third Circuit's determination, the Commission overturned the Board's decision to admit a NEPA terrorism contention in the Diablo Canyon License Renewal proceeding, a facility located in the Ninth Circuit. *Pacific Gas & Electric Co.* (Diablo Canyon Nuclear Power Plant), CLI-11-11, 74 NRC (slip op. at 40) (2011). The Commission reaffirmed that "the staff's determination in the GEIS that the environmental impacts of a terrorist attack were bounded by those resulting from internally-initiated events, was sufficient to address the environmental impacts of terrorism." *Id.*

In sum, the Commission has found that the issuance of a facility license is not the "proximate cause" of a terrorist attack at that facility. Thus, it is not required to prepare an EIS discussion on the potential impacts of a terrorist attack. However, due to the decision of the Ninth Circuit, the NRC will prepare an analysis of the environmental impacts of a terrorist attack for licensing actions of facilities within the geographical boundaries of the Ninth circuit. In addition, the Third Circuit has held that the GEIS for License Renewal constitutes such an analysis for license renewals.

E.3.1 Impact of New Information on Accidents Initiated by Internal Events

With few exceptions, the severe accident analyses formulating the basis for the 1996 GEIS were limited to consideration of reactor accidents caused by internal events. The GEIS addressed the impacts from external events qualitatively, and external events are covered in more detail in Section E.3.2 of this revision. The impacts from the 1996 GEIS were based on the original license EISs for the 28 nuclear power plant sites listed in Table 5.1 of the GEIS. The source terms and their likelihood used in the plant-specific original EISs to calculate the airborne pathway environmental impacts of accidents were, in turn, usually based upon information contained in NUREG-0773 (NRC 1982). NUREG-0773 is an update of the original Reactor Safety Study (NRC 1975). These source terms and frequencies were used along with site-specific meteorology, population distributions, and emergency planning characteristics to calculate the airborne pathway environmental impacts. These EISs were issued in the 1981 to 1986 time frame. Thus, while the GEIS was published in 1996, it was primarily based on information from the 1980s.

Since the publication of NUREG-0773, many additional studies have been completed on the likelihood and consequences of reactor accidents initiated by internal events at full power. These studies include NUREG-1150 (NRC 1990b), NUREG/CR-5305 (NRC 1992), and licensee responses to Generic Letter 88-20, Supplement 1 (i.e., the IPE program). Licensees have further developed their IPE-vintage probabilistic risk assessment (PRA) models to support risk-informed licensing actions, including license renewal SAMA analysis. In addition, the NRC has developed standardized plant analysis risk (SPAR) models for all operating plants which can be used to calculate core damage frequencies (CDFs) for internal events.

The purpose of this section is to assess how results from more up-to-date internal event information compare to those on which the 1996 GEIS was based. The evaluation contained in this section compares the CDFs that formed the basis for the 1996 GEIS, and offsite doses directly from the 1996 GEIS, to the newer information. The comparison is done for pressurized water reactors (PWRs) and boiling water reactors (BWRs) and covers each of the plants listed in Table 5.1 of the 1996 GEIS. Changes in source terms (i.e., the quantity, form, and timing of radioactive material released to the environment) are assessed in Section E.3.3.

E.3.1.1 Airborne Pathway Impacts

As a first step in the comparison, the CDFs from the original EISs are compared to the CDFs reported in the plant-specific IPEs for the PWRs and BWRs considered by the 1996 GEIS. Tables E-1 and E-2 show these comparisons. As can be seen in Tables E-1 and E-2, for many plants, the IPE CDFs are smaller than those from the original EISs, particularly for BWRs. The mean of the IPE CDFs listed in Tables E-1 and E-2 are lower than the corresponding mean EIS CDF by 30 percent for PWRs and by more than a factor of 3 for BWRs. Accordingly, the likelihood of an accident that leads to core damage would be comparable to or less for PWRs, and significantly less for BWRs, than that used as the basis for the 1996 GEIS.

Additional comparisons can be made using information from NUREG-0773 (NRC 1982), the original EISs, NUREG-1150 (NRC 1990b), the IPEs, NUREG/CR-5305 (NRC 1992), recent analysis using SPAR models, and license renewal applications received to date. These comparisons are shown in Table E-3. In general, the Level 1 (CDF) results are comparable to or less than the corresponding Level 1 information from the GEIS. Furthermore, the newer estimates (license renewal and SPAR) are up to a factor of 2.5 lower than the mean IPE CDFs from Tables E-1 and E-2.

The comparison of Level 3 (offsite consequences) information is made difficult due to differences in the values reported between older and newer assessments. Older assessments tended to provide mean and/or upper bound population doses for the entire region surrounding the nuclear power plant (as far as 1000 mi). Newer assessments tend to provide mean values within 50 mi. NUREG-1150 provided distributions for both within 50 mi and the site region and is used as a bridge in this comparison.

The mean of population dose results from the original EISs of the 28 sites that considered severe accidents are a factor of 2 to 4 lower than the mean of the plant-specific upper bound estimates used in the 1996 GEIS for those same 28 sites. The mean population doses from NUREG-1150 (site region results) are, in turn, a factor of 10 to 100 less than the original EIS mean value. In actuality, the difference is even larger, because the NUREG-1150 estimate covers a larger area (site region for NUREG-1150 versus 150 mi for the EISs). The NUREG-1150 results for a 50-mi radius are a factor of 4 to 10 lower than the site region results. The mean of license renewal results (for a 50-mile region) are somewhat higher than the mean results reported in NUREG-1150 for a 50-mile region, but are still well below the population dose values reported in the original environmental impact statements for the 28 sites and used in the 1996 GEIS.

Table E-1. PWR Internal Event (Full Power) Comparison

Plant	Original EIS Estimated CDF[(a)]	IPE CDF[(b)]
Beaver Valley 2	1.0×10^{-4}/yr	1.9×10^{-4}/yr
Braidwood 1, 2	1.0×10^{-4}/yr	2.7×10^{-5}/yr
Byron 1, 2	4.8×10^{-5}/yr	3.1×10^{-5}/yr
Callaway 1	4.8×10^{-5}/yr	5.9×10^{-5}/yr
Catawba 1, 2	4.8×10^{-5}/yr	5.8×10^{-5}/yr
Comanche Peak 1, 2	4.8×10^{-5}/yr	5.7×10^{-5}/yr
Shearon Harris 1	4.8×10^{-5}/yr	7.0×10^{-5}/yr
Indian Point 2, 3	3.5×10^{-4}/yr, 3.4×10^{-4}/yr	3.1×10^{-5}/yr, 4.4×10^{-5}/yr
Millstone 3	2.0×10^{-4}/yr	5.6×10^{-5}/yr
Palo Verde 1, 2, 3	4.8×10^{-5}/yr	9.0×10^{-5}/yr
San Onofre 2, 3	4.8×10^{-5}/yr	3.0×10^{-5}/yr
Seabrook 1	4.8×10^{-5}/yr	6.1×10^{-5}/yr[(c)]
South Texas 1, 2	4.4×10^{-5}/yr	4.3×10^{-5}/yr
St. Lucie 2	4.8×10^{-5}/yr	2.6×10^{-5}/yr
Summer 1	4.9×10^{-5}/yr	2.0×10^{-4}/yr
Vogtle 1, 2	1.0×10^{-4}/yr	4.9×10^{-5}/yr
Waterford 3	4.8×10^{-5}/yr	1.8×10^{-5}/yr
Wolf Creek 1	4.8×10^{-5}/yr	4.2×10^{-5}/yr
Mean value	8.4×10^{-5}/yr	5.9×10^{-5}/yr
Median value	4.8×10^{-5}/yr	4.9×10^{-5}/yr

(a) Obtained by summing individual atmospheric release sequences, including intact containment sequences.
(b) Source: NRC 2003, unless otherwise noted.
(c) Obtained from the licensee's IPEEE submittal.

To summarize, the general contribution to decreased estimated doses are a factor of 2 to 4 simply due to the conservatism built into the 1996 GEIS values. An additional decrease in estimated doses of 10 to 100 is seen when comparing the EIS results to the NUREG-1150 results and a factor of 5 to 33 when comparing the EIS results to license renewal SAMA results.

E.3.1.2 Other Pathway Impacts

Any change in the likelihood of accidents that release substantial amounts of radioactive material to the environment not only affects the airborne pathway, but also the surface water and groundwater pathways and the resulting economic impacts from any pathway. The information in Tables E-1, E-2, and E-3 indicate that the likelihood and impacts of airborne pathway releases is smaller than that used in the 1996 GEIS. Since this pathway directly

Table E-2. BWR Internal Event (Full Power) Comparison

Plant	Original EIS Estimated CDF[(a)]	IPE CDF[(b)]
Clinton 1	2.4×10^{-5}/yr	2.7×10^{-5}/yr
Fermi 2	2.4×10^{-5}/yr	5.7×10^{-6}/yr
Grand Gulf 1	2.4×10^{-5}/yr	1.7×10^{-5}/yr
Hope Creek	1.0×10^{-4}/yr	4.6×10^{-5}/yr
Limerick 1 ,2	8.9×10^{-5}/yr	4.3×10^{-6}/yr
Nine Mile Point 2	1.1×10^{-4}/yr	3.1×10^{-5}/yr
Perry 1	2.4×10^{-5}/yr	1.3×10^{-5}/yr
River Bend	9.5×10^{-5}/yr	1.6×10^{-5}/yr
Susquehanna 1, 2	2.4×10^{-5}/yr	5.6×10^{-7}/yr[(c)]
WNP-2[(d)]	2.4×10^{-5}/yr	1.8×10^{-5}/yr
Mean value	5.4×10^{-5}/yr	1.5×10^{-5}/yr
Median value	2.4×10^{-5}/yr	1.45×10^{-5}/yr

(a) Obtained by summing individual atmospheric release sequences, including intact containment sequences.
(b) Source: NRC 2003, unless otherwise noted.
(c) Revised 1998 IPE; obtained from NUREG-1437, Supp. 35, Appendix G.
(d) WNP-2 = Washington Nuclear Project 2 (i.e., Columbia).

affects the surface water pathway, it is reasonable to conclude that the likelihood of the surface pathway impacts would also be smaller and would continue to be bounded by the airborne pathway. The decreased likelihood of any pathway impacts would indicate the reduced likelihood of any subsequent economic impacts. This assumption is consistent with the results of the 1996 GEIS.

Furthermore, some information is available regarding basemat melt-through sequences, which could impact the groundwater pathway:

- WASH-1400 (NRC 1975) used a frequency of 4×10^{-5}/yr for basemat melt-through sequences;
- NUREG-0773 (NRC 1982) used a generic frequency of 3×10^{-5}/yr and a site-specific frequency of 1.1×10^{-5}/yr for Indian Point Units 2 and 3;
- NUREG-1150 (NRC 1990b) calculated the basemat melt-through frequencies for Surry and Sequoyah to be 2.4×10^{-6}/yr and 1×10^{-5}/yr, respectively;

Table E-3. Comparisons with Other Risk Information (Full Power Internal Events)

Reactor Type	Comparison	Information Source	CDF (mean/point estimate)	Person-Rem per Year (Mean, except as noted)	
				Region[a]	50-mi
PWR	GEIS Basis				
		NUREG-0773[b]	6×10^{-5}/yr		
		Original EIS[c]	8.4×10^{-5}/yr	932	
		1996 GEIS[c]		2,200[d]	
	Update	NUREG-1150 Plants			
		- Surry	4×10^{-5}/yr	~30	~6
		- Sequoyah	5.6×10^{-5}/yr	~80	~10
		IPE			
		- Catawba	5.8×10^{-5}/yr		15.66
		- McGuire	4×10^{-5}/yr		4.6
		- Surry	1.25×10^{-4}/yr		
		- Sequoyah	1.7×10^{-4}/yr		
		License Renewal[e]	3.9×10^{-5}/yr		18.1
		SPAR (v3.45)[c]	2.3×10^{-5}/yr		
BWR	GEIS Basis	NUREG-0773[b]	2×10^{-5}/yr		
		Original EIS[c]	5.4×10^{-5}/yr	577	
		1996 GEIS[c]		2,720[d]	
	Update	NUREG-1150 Plants			
		- Grand Gulf	4×10^{-6}/yr	~5	~0.5
		- Peach Bottom	4.4×10^{-6}/yr	~30	~7
		NUREG/CR-5305			
		- LaSalle	4×10^{-5}/yr	1,500[f]	66[e]
		IPE			
		- Peach Bottom	5.5×10^{-6}/yr		
		- LaSalle	4.7×10^{-5}/yr		
		- Grand Gulf	1.7×10^{-5}/yr		

Reactor Type	Comparison	Information Source	CDF (mean/point estimate)	Person-Rem per Year (Mean, except as noted)	
				Region[a]	50-mi
		License Renewal[e]	1.4×10^{-5}/yr		14.5
		SPAR (v3.45)[b]	8×10^{-6}/yr		

(a) For the EISs and GEIS, the employed distance is 150 mi; for NUREG-1150 and NUREG/CR-5305, the employed distance is 1,000 mi.

(b) Based on Table 22 (CDF) of that document; PWR CDF cited is for Surry and BWR corresponds to Peach Bottom.

(c) Values are for those plants listed in Tables E-1 and E-2.

(d) Note that this is the mean of the distribution of 95th percent UCB values.

(e) Mean values for all plants that have applied for license renewal as of August 2008; in a few cases (Beaver Valley, Calvert Cliffs, Ginna, and Nine Mile Point), the site-specific population dose values used included both internal and external events.

(f) Includes both internal and external events.

- A sample of IPE results showed basemat melt-through frequencies ranging from 1×10^{-6}/yr to 4×10^{-6}/yr; and

- A sample of license renewal application results showed basemat melt-through frequencies ranging from 2×10^{-7}/yr to 6×10^{-6}/yr.

For the 1996 GEIS, a conservative value of 1×10^{-4}/yr was used (see Section 5.3.3.4 of the 1996 GEIS), which is higher than any of the values cited above. As such, it is concluded that the basemat melt-through frequencies used in the 1996 GEIS to assess the groundwater pathway are bounding.

For BWRs, no quantitative basemat melt-through information was available. It is expected that for BWRs, containment failure by overpressure would occur before basemat melt-though. In addition, if basemat melt-through sequences do occur, their frequency would be less than that for PWRs due to the lower CDFs for BWRs.

E.3.1.3 Conclusion

The PWR and BWR accident frequencies that form the basis for the environmental impacts shown in the 1996 GEIS are, in most cases, comparable to or higher than the updated accident frequencies shown in Tables E-1, E-2, and E-3. In addition, the population dose estimates presented in Table E-3 demonstrate the conservatism in the 1996 GEIS values, both from the standpoint of reduced population dose from more recent estimates and the conservatism built into the GEIS methodology.

E.3.2 Impact of Accidents Initiated by External Events

The 1996 GEIS included a qualitative assessment of the environmental impacts of accidents initiated by external events (see Section 5.3.3.1 of that document). The purpose of this section is to consider updated information regarding potential external event impacts. The sources of information used in this assessment are (1) NUREG-1150 (NRC 1990b) (and the supporting documentation in NUREG/CR-4551 [NRC 1990a]), which assessed seismic and fire events for two plants (Surry and Peach Bottom); (2) NUREG/CR-5305 (NRC 1992), which analyzed the risk from seismic and fire events for one plant (LaSalle); and (3) the results from the IPEEE program, as documented in NUREG-1742 (NRC 2003). The IPEEE program was initiated in the early 1990s and required all operating plants in the United States to do an assessment to identify vulnerabilities to severe accidents initiated by external events and report the results to the NRC, along with any identified improvements and/or corrective actions. NUREG-1742 documents the perspectives derived from the technical reviews of the IPEEE results.

Typically, the external events that contribute the most to plant risk are seismic and fires. In some cases, high winds, floods, and tornados may contribute to plant risk; however, these contributions are generally much lower than those from seismic and fire events. Therefore, the assessment of the environmental impact from external events provided here focuses on seismic and fire events. This is consistent with the results obtained from the IPEEEs and the perspectives articulated in NUREG-1742.

E.3.2.1 Airborne Pathway Impacts

The assessment in this section is based upon a comparison of the risks and environmental impacts from severe accidents initiated by external events to those initiated by internal events, based on the aforementioned information sources.

Level 1 Comparison (CDF)

From the IPEEE the following insights can be drawn:

(1) For a majority of plants, fire and/or seismic events are important contributors to risk.

(2) The contributions to CDFs from fire events are comparable to the contribution to CDFs from internal events. The IPEEE CDF values for fire-initiated events are shown in Tables E-4 and E-5 along with the IPE internal event CDFs. For the plants listed in Tables E-4, the PWR fire CDF is about half the internal event CDF. For the BWR plants in Table E-5, the fire CDF is roughly 50 percent higher than the internal events CDF. Section 3.3.1.1 of NUREG-1742 (NRC 2003) provides a comparison of fire and internal events for the entire fleet of plants, and similarly concludes that BWR results are comparable, while PWR results are slightly lower for fire CDF.

However, the IPEEE fire event CDFs are much lower than the internal event CDFs from the original EISs (basis for the 1996 GEIS). The mean value of the PWR fire event CDFs in Table E-4 is one-third of the PWR internal event CDF from the EISs (see Table E-1), and the mean value of the BWR fire event CDFs in Table E-5 is less than half the BWR internal event CDF from the original EISs (see Table E-2).

(3) The contributions to CDF from seismic events are comparable to the contribution from internal events. For plants listed in Tables E-1 and E-2 that reported seismic CDFs as part of their IPEEE submittals, these CDFs are contained in Tables E-4 and E-5. Although sparse, these values suggest seismic CDFs are lower than or comparable to internal event CDFs. Section 2.6.1 of NUREG-1742 considers all reporting plants, and states that the largest group of reported seismic CDFs were in the range of 1×10^{-5} to 1×10^{-4} (same order of magnitude as the basis for the 1996 GEIS), with the next largest

Table E-4. PWR Internal, Fire, and Seismic Event CDF Comparison (Full Power)[a]

Plant	IPE Internal Events CDF	IPEEE Fire CDF	IPEEE Seismic CDF (EPRI/Other/Update)	IPEEE Seismic CDF (LLNL)
Beaver Valley 2	1.9×10^{-4}/yr	1.1×10^{-5}/yr	1×10^{-5}/yr	2.3×10^{-5}/yr
Braidwood 1, 2	2.7×10^{-5}/yr	3.9×10^{-6}/yr 3.8×10^{-6}/yr		
Byron 1, 2	3.1×10^{-5}/yr	4.2×10^{-6}/yr 5.3×10^{-6}/yr		
Callaway 1	5.9×10^{-5}/yr	8.9×10^{-6}/yr		
Catawba 1, 2	5.8×10^{-5}/yr	4.6×10^{-6}/yr	1.6×10^{-5}/yr	
Comanche Peak 1, 2	5.7×10^{-5}/yr	2.1×10^{-5}/yr		
Shearon Harris 1	7.0×10^{-5}/yr	1.3×10^{-5}/yr		
Indian Point 2, 3	3.1×10^{-5}/yr 4.4×10^{-5}/yr	1.8×10^{-5}/yr 5.6×10^{-5}/yr	1.3×10^{-5}/yr 5.9×10^{-5}/yr	1.5×10^{-5}/yr 4.4×10^{-5}/yr
Millstone 3	5.6×10^{-5}/yr	4.8×10^{-6}/yr	9.1×10^{-6}/yr	
Palo Verde 1, 2, 3	9.0×10^{-5}/yr	8.7×10^{-5}/yr		
San Onofre 2, 3	3.0×10^{-5}/yr	1.6×10^{-5}/yr	1.7×10^{-5}/yr	
Seabrook 1	6.1×10^{-5}/yr[b]	1.2×10^{-5}/yr	1.2×10^{-5}/yr	1.3×10^{-4}/yr
South Texas 1, 2	4.3×10^{-5}/yr	5.1×10^{-7}/yr	1.9×10^{-7}/yr	2.2×10^{-5}/yr
St. Lucie 2	2.6×10^{-5}/yr	1.9×10^{-4}/yr		
Summer 1	2.0×10^{-4}/yr	8.5×10^{-5}/yr		
Vogtle 1, 2	4.9×10^{-5}/yr	1.0×10^{-5}/yr		
Waterford 3	1.8×10^{-5}/yr	7.0×10^{-6}/yr		
Wolf Creek 1	4.2×10^{-5}/yr	7.6×10^{-6}/yr		
Mean Value	5.9×10^{-5}/yr	2.8×10^{-5}/yr	1.5×10^{-5}/yr	4.3×10^{-5}/yr

(a) Source: NRC 2003, unless otherwise stated.
(b) Obtained from the licensee's IPEEE submittal.

group being 1×10^{-6} to 1×10^{-5} (one order of magnitude lower than the basis for the 1996 GEIS).

(4) As a result of the IPEEE program, most licensees have made improvements to plant hardware, procedures, or training programs. Although not generally quantified as part of the IPEEE, those improvements are, in many cases, considered to have lowered the reported risk estimates.

Table E-6 compares CDFs from NUREG-1150 (NRC 1990b) and NUREG/CR-5305 (NRC 1992) for internal, fire, and seismic events with the internal events from the original EISs (which

Table E-5. BWR Internal, Fire, and Seismic Event CDF Comparison (Full Power)[a]

Plant	IPE Internal Events CDF	IPEEE Fire CDF	IPEEE Seismic CDF (EPRI/Other/Update)	IPEEE Seismic CDF (LLNL)
Clinton 1	2.7×10^{-5}/yr	3.6×10^{-6}/yr		
Fermi 2	5.7×10^{-6}/yr	2.2×10^{-5}/yr		
Grand Gulf 1	1.7×10^{-5}/yr	8.9×10^{-6}/yr		
Hope Creek	4.6×10^{-5}/yr	8.1×10^{-5}/yr	1.1×10^{-6}/yr	3.6×10^{-6}/yr
Limerick 1, 2	4.3×10^{-6}/yr	NA[b]		
Nine Mile Point 2	3.1×10^{-5}/yr	1.4×10^{-6}/yr	2.5×10^{-7}/yr	1.2×10^{-6}/yr
Perry 1	1.3×10^{-5}/yr	3.3×10^{-5}/yr		
River Bend	1.6×10^{-5}/yr	2.3×10^{-5}/yr		
Susquehanna 1, 2	5.6×10^{-7}/yr[c]	3.6×10^{-8}/yr		
WNP-2[d]	1.8×10^{-5}/yr	5.5×10^{-5}/yr	2.1×10^{-5}/yr	
Mean Value	1.5×10^{-5}/yr	2.3×10^{-5}/yr	7.5×10^{-6}/yr	2.4×10^{-6}/yr

(a) Source: NRC 2003, unless otherwise stated.
(b) NA = not available.
(c) Revised 1998 IPE; obtained from NUREG-1437, Supp. 35, Appendix G.
(d) WNP-2 = Washington Nuclear Project 2 (i.e., Columbia).

Table E-6. NUREG-1150 and NUREG/CR-5305 Fire and Seismic CDFs

Plant	Internal Events (mean value)	Fire Events (mean value)	Seismic Events (mean value)[a]	1996 GEIS Basis Internal Events (mean value)
Surry (NUREG-1150)	4×10^{-5}/yr	1.1×10^{-5}/yr	1.9×10^{-4}/yr	8.4×10^{-5}/yr[b]
Peach Bottom (NUREG-1150)	4.4×10^{-6}/yr	2×10^{-5}/yr	7.5×10^{-5}/yr	5.4×10^{-5}/yr[c]
LaSalle (NUREG/CR-5305)	4×10^{-5}/yr	5.5×10^{-5}/yr	8×10^{-7}/yr	5.4×10^{-5}/yr[c]

(a) Based on the LLNL seismic hazard distribution results.
(b) This value is the mean of the CDFs of all PWRs listed in Table 5.1 of the 1996 GEIS.
(c) This value is the mean of the CDFs of all BWRs listed in Table 5.1 of the 1996 GEIS.

formed the basis for the 1996 GEIS). As can be seen in this table, the NUREG-1150 and NUREG/CR-5305 fire and seismic CDFs are comparable to those supporting the 1996 GEIS, with a number of both relatively lower and higher comparisons.[j]

In support of early site permits for new reactors, the NRC staff reviewed updates to seismic source and ground motion models provided by applicants. The updates to seismic data and models could result in estimated seismic hazard levels at some current central and eastern U.S. operating sites that would be higher than seismic hazard values used in design and previous evaluations (such as the IPEEEs). Due to its relevance for other licensing actions, the issue is being pursued as part of the Generic Issues Program, as Generic Issue 199 (GI-199). A preliminary assessment performed for the affected plants as part of GI-199 indicates that the average increase in seismic CDF relative to the IPEEE-era estimates would be about 1×10^{-5} per year. However, this assessment also indicates that on average, the updated seismic CDF remains slightly (approximately 30 percent) less than the internal events CDF.

Level 3 Comparison (Offsite Consequences)

To obtain quantitative information on the airborne pathway environmental impacts of severe accidents caused by external events, IPEEE, NUREG-1150, and NUREG/CR-5305 results can be used to compare against the internal event airborne pathway impacts contained in the 1996 GEIS. The following discussion summarizes the airborne pathway environmental impact information available.

The IPEEE provided external event environmental impact information (i.e., early fatalities, latent fatalities, and population dose) for Catawba and McGuire. This information showed the impacts of external events to be much less (i.e., one to two orders of magnitude) than those estimated for internally initiated events at full power in the 1996 GEIS for Catawba and McGuire (see Table E-7). Recall that while this is a comparison of mean values versus 95 percent upper confidence bound (UCB) values, the 95 percent UCB values are the ones used for the basis of the 1996 GEIS. Thus, this comparison shows that more realistic estimates are significantly lower than the conservative estimates used in the GEIS.

(j) The NUREG-1150 values represented best-estimate values at the time they were completed. For Surry, the Lawrence Livermore National Laboratory (LLNL) NUREG-1150 curve is uniformly higher than other seismic hazard estimates (e.g., the Electric Power Research Institute [EPRI] and LLNL curves used for the IPEEEs, recent United States Geological Survey curves). For Peach Bottom, the EPRI NUREG-1150 curve is uniformly lower.

Table E-7. Catawba and McGuire Results for Internal and External Events

Impact	Catawba External Events	Catawba Internal Events	Catawba 1996 GEIS Internal Events (95 percent UCB)	McGuire External Events	McGuire Internal Events	McGuire 1996 GEIS Internal Events (95 percent UCB)
Total person-rem per year	43.6	15.6	1,880	10.7	4.6	1,806
Total early fatality risk	7.8×10^{-6}/yr	5.9×10^{-6}/yr	1.7×10^{-2}/yr	2.2×10^{-6}/yr	8.2×10^{-7}/yr	1.0×10^{-2}/yr
Total latent fatality risk	2.7×10^{-3}/yr	9.4×10^{-4}/yr	1.4/yr[(a)]	7.4×10^{-4}/yr	3.2×10^{-4}/yr	1.4/yr[(a)]

(a) These values include the factor of 10 adjustment made in the 1996 GEIS (see Section 5.3.3.2.3 of the 1996 GEIS).

Fire Events

NUREG-1150 provides quantitative information on the airborne pathway environmental impact from fires for Surry and Peach Bottom. This information is shown in Tables E-8 and E-9 along with the full power, internal event environmental impact information from NUREG-1150 and the 1996 GEIS. NUREG/CR-5305 provides similar information for LaSalle, as presented in Table E-10. Tables E-8 through E-10 present 95th percentile results for all values. As can be seen from these tables, even 95th percentile values from NUREG-1150 and NUREG/CR-5305 are significantly lower (at least by 1 order of magnitude) than the conservative values used in the 1996 GEIS.

Seismic Events

Table E-11 presents mean results from the second-tier NUREG-1150 study documentation (NUREG/CR-4551; NRC 1990a) for impacts due to seismic initiators at Surry and Peach Bottom. As can be seen from this table, the mean results from the NUREG-1150 study are, in most cases, significantly smaller than the 95th percentile estimates used in the 1996 GEIS.

E.3.2.2 Other Pathway Impacts

With respect to the other pathways (open bodies of water and groundwater), the IPEEE, NUREG-1150, and NUREG/CR-5305 analysis did not address their impacts on human health. The 1996 GEIS estimated these impacts for reactor accidents from full power (internal events only) using the results from site-specific information on surface water and groundwater areas, volumes, flow-rates, and geology to assess contamination of water by comparing the site-specific information to that used in NUREG-0440 (NRC 1978), which assessed the contamination of surface water and groundwater from reactor accidents.

Table E-8. Impacts of Accidents Caused by Fire Events (Surry)

Impact	NUREG-1150 Fire Events (95th percentile)	NUREG-1150 Internal Events (95th percentile)	1996 GEIS Internal Events (95th percentile)
Individual risk			
- EF[(a)] (1 mi)	~1.5 × 10^{-10}/yr	~5 × 10^{-8}/yr	Not available
- LF[(b)] (10 mi)	~1.5 × 10^{-10}/yr	~1 × 10^{-8}/yr	Not available
Total person-rem per year (entire region)	~2	~150	1,200
Total early fatality risk	~1 × 10^{-8}/yr	~4 × 10^{-6}/yr	1.6 × 10^{-2}/yr
Total latent fatality risk	~6 × 10^{-4}/yr	~3 × 10^{-2}/yr	0.9/yr

(a) EF = early fatality risk. The individual early fatality risk within one mile (1.6 km) is the frequency (per year) that a person living within one mile (1.6 km) of the site boundary will die within a year due to the accident. The entire population within one mile is considered to obtain an average value.

(b) LF = latent fatality risk. The individual latent cancer fatality risk within 10 miles (16 km) is the frequency (per year) that a person living within 10 miles (16 km) of the plant will die many years later from cancer due to radiation exposure received from the accident. The entire population within 10 miles (16 km) is considered to obtain an average value.

Table E-9. Impacts of Accidents Caused by Fire Events (Peach Bottom)

Impact	NUREG-1150 Fire Events (95th percentile)	NUREG-1150 Internal Events (95th percentile)	1996 GEIS Internal Events (95th percentile)
Individual risk			
- EF[(a)] (1 mi)	~1.5 × 10^{-9}/yr	~2.5 × 10^{-10}/yr	Not available
- LF[(b)] (10 mi)	~1 × 10^{-8}/yr	~1.5 × 10^{-9}/yr	Not available
Total person-rem per year (entire region)	~700	~100	2,950
Total early fatality risk	~1.5 × 10^{-6}/yr	~1 × 10^{-7}/yr	4.2 × 10^{-3}/yr
Total latent fatality risk	~0.15/yr	~2 × 10^{-2}/yr	2.0/yr

(a) EF = early fatality risk. The individual early fatality risk within one mile (1.6 km) is the frequency (per year) that a person living within one mile (1.6 km) of the site boundary will die within a year due to the accident. The entire population within one mile is considered to obtain an average value.

(b) LF = latent fatality risk. The individual latent cancer fatality risk within 10 miles (16 km) is the frequency (per year) that a person living within 10 miles (16 km) of the plant will die many years later from cancer due to radiation exposure received from the accident. The entire population within 10 miles (16 km) is considered to obtain an average value.

Table E-10. Impacts of Accidents Caused by Fire Events (LaSalle)

Impact	NUREG/CR-5305 Fire Events (95th percentile)	NUREG/CR-5305 Internal Events (95th percentile)	1996 GEIS Internal Events (95th percentile)
Individual risk			
- EF[a] (1 mi)	~1.1 × 10^{-10}/yr	~1.5 × 10^{-10}/yr	Not available
- LF[b] (10 mi)	~1.0 × 10^{-8}/yr	~1.3 × 10^{-8}/yr	Not available
Total person-rem per year	~1,920	~2,600	2,898
Total early fatality risk	~9 × 10^{-9}/yr	~1.2 × 10^{-8}/yr	3.6 × 10^{-3}/yr
Total latent fatality risk	~0.3/yr	~0.4/yr	2.0/yr

(a) EF = early fatality risk. The individual early fatality risk within one mile (1.6 km) is the frequency (per year) that a person living within one mile (1.6 km) of the site boundary will die within a year due to the accident. The entire population within one mile is considered to obtain an average value.

(b) LF = latent fatality risk. The individual latent cancer fatality risk within 10 miles is the frequency (per year) that a person living within 10 miles (16 km) of the plant will die many years later from cancer due to radiation exposure received from the accident. The entire population within 10 miles (16 km) is considered to obtain an average value.

Table E-11. Impacts of Accidents Caused by Seismic Events

	Surry		Peach Bottom	
Impact	NUREG/CR-4551 Surry[a] LLNL (*EPRI*) Hazard Curve	1996 GEIS (95th percentile)	NUREG/CR-4551 Peach Bottom[a] LLNL (*EPRI*) Hazard Curve	1996 GEIS (95th percentile)
Individual risk				
- EF[b] (1 mi)	1.8 × 10^{-7}/yr (*1.8 × 10^{-8}/yr*)		1.6 × 10^{-6}/yr (*5.3 × 10^{-8}/yr*)	
- LF[c] (10 mi)	3.1 × 10^{-8}/yr (*3.8 × 10^{-9}/yr*)		3.4 × 10^{-7}/yr (*1.1 × 10^{-8}/yr*)	
Total person-rem per year	45 (*6.7*)	1,200	460 (*17*)	2,950
Total early fatality risk	9.3 × 10^{-5}/yr (*1.4 × 10^{-5}/yr*)	1.6 × 10^{-2}/yr	3.0 × 10^{-3}/yr (*8.8 × 10^{-5}/yr*)	4.2 × 10^{-3}/yr
Total latent fatality risk	3.9 × 10^{-2}/yr (*5.6 × 10^{-3}/yr*)	0.9/yr	2.5 × 10^{-1}/yr (*9.9 × 10^{-3}/yr*)	2.0/yr

(a) Mean values.

(b) EF = early fatality risk. The individual early fatality risk within one mile (1.6 km) is the frequency (per year) that a person living within one mile (1.6 km) of the site boundary will die within a year due to the accident. The entire population within one mile is considered to obtain an average value.

(c) LF = latent fatality risk. The individual latent cancer fatality risk within 10 miles (16 km) is the frequency (per year) that a person living within 10 miles (16 km) of the plant will die many years later from cancer due to radiation exposure received from the accident. The entire population within 10 miles (16 km) is considered to obtain an average value.

With the airborne pathway impacts from external events much less than the internal event airborne pathway impacts in the 1996 GEIS, it is reasonable to conclude that the impact of accidents caused by external events on surface water and groundwater contamination will also be much less than the impacts contained in the 1996 GEIS. Due to the longer time before the population is exposed and the effects of interdiction of contaminated food, only latent fatalities are expected to result from these pathways. Therefore, the environmental impacts of surface and groundwater contamination caused by accidents initiated by external events are bounded by the impacts stated in the 1996 GEIS. This same conclusion can also be drawn with respect to the economic impacts that are caused by the environmental contamination.

E.3.2.3 Conclusion

In summary, it is concluded that the CDFs from severe accidents initiated by external events, as quantified in NUREG-1150 (NRC 1990b) and the other sources cited above, are comparable to those from accidents initiated by internal events but lower than the CDFs that formed the basis for the 1996 GEIS. The environmental impacts from externally initiated events are generally significantly lower (one or more orders of magnitude) than those used in the 1996 GEIS.

E.3.3 Impact of New Source Term Information

The 1996 GEIS used information from 28 plant-specific EISs to project the environmental impact from all 118 plants analyzed (see Table 5.5 in the 1996 GEIS). The 28 sites chosen were those for which the impacts from severe accidents were analyzed in their plant-specific EISs. As stated in Section 5.3.3.1 of the 1996 GEIS, the source terms (i.e., the magnitude, timing, and characteristics of the radioactive material released to the environment) used in the EIS analyses for the 28 sites (and subsequently used to estimate the environmental impacts from all plants) were generally based on those documented in NUREG-0773 (NRC 1982). The NUREG-0773 source terms represented an update (re-baseline) of the source terms used in WASH-1400 (NRC 1975). The source terms in NUREG-0773 were developed for PWRs and BWRs and are shown in Tables 13 and 14A of that document. NUREG-0773 states that the provided source terms are based on models that have "known deficiencies which would tend to give overestimates of the magnitude of the releases."

Since completion of NUREG-0773, additional information on source terms has been developed through experimental and analytical programs. The purpose of this section is to assess the impact of new source term information on the environmental impacts described in the 1996 GEIS. The new source term information assessed is that used in NUREG-1150 (NRC 1990b) as updated and simplified in NUREG/CR-6295 (NRC 1997b).

E.3.3.1 Airborne Pathway Impact

Tables E-12 and E-13 present a comparison of the results for large release sequences from NUREG-0773 (NRC 1982) and NUREG/CR-6295 (NRC 1997b). These sequences typically dominate the total risk from all severe accidents. In this case, large release sequences have been selected from the full set of sequences in each study based on a total iodine release fraction of 10 percent or higher. These tables present release frequencies, timings, and release fractions for iodine and cesium, which are the elements that contribute the most to early (iodine) and latent (cesium) fatalities. Only limited comparisons between the studies are possible due to differences in the sequences analyzed in each study and their associated release modes. Nevertheless the following observations can be made:

- The sum of the release frequencies from NUREG/CR-6295 is lower than those from NUREG-0773 for all containment types, with the exception of the NUREG/CR-6295 LaSalle sequences. However, the higher release frequency for LaSalle is offset by lower release fractions at LaSalle.

Table E-12. NUREG-0773 and NUREG/CR-6295 Large Source Terms (PWRs)

Source		Sequence	Frequency	Release Time (hr)	Release Duration (hr)	Post Core Uncovery Delta (hr)[a]	Iodine Release Fraction	Cesium Release Fraction
NUREG-0773	Surry	Event V (Bypass)	4×10^{-6}/yr	1	1	0.5	0.64	0.82
		TMLB'-δ (CF during CD)	3×10^{-6}/yr	2.5	0.5	1	0.31	0.39
		PWR-3 (CR during CD)	3×10^{-6}/yr	5	1.5	2	0.2	0.2
		Sum	1×10^{-5}/yr					
NUREG/CR-6295	Surry	RSUR1[b] (CF at VB)	2.9×10^{-7}/yr	6	2	1	0.35	0.31
		RSUR4[b] (Bypass)	1.6×10^{-6}/yr	1	2.5	0.7	0.12	0.12
		Sum	1.9×10^{-6}/yr					
	Sequoyah	RSEQ1[b] (CF during CD)	2.8×10^{-7}/yr	5.5	2	0.5	0.59	0.62
		RSEQ2[b] (CF at VB)	3.6×10^{-6}/yr	6	2	1	0.18	0.19
		RSEQ5[b] (Bypass)	3.1×10^{-6}/yr	1	2.5	0.7	0.12	0.12
		Sum	7×10^{-6}/yr					

(a) For NUREG-0773, this represents the interval of time between the decision to take protective measures and the start of the release; for NUREG/CR-6295, this represents the time between core uncovery and the start of the release.
(b) These source terms have multiple plumes, which have been summed here for ease of comparison.
Bypass = fission product released from the reactor bypass the containment.
CF = containment failure.
CD = core damage.
VB = reactor vessel branch.

Table E-13. NUREG-0773 and NUREG/CR-6295 Large Source Terms (BWRs)

Source		Sequence	Frequency	Release Time (hr)	Release Duration (hr)[(a)]	Post Core Uncovery Delta (hr)[(a)]	Iodine Release Fraction	Cesium Release Fraction
NUREG-0773	Peach Bottom	AEα' (CF before VB)	2×10^{-9}/yr	0.8	0.5	0.5	0.3	0.6
		AEα (CF before VB, scrub)	1×10^{-9}/yr	0.8	0.5	0.5	0.2	0.4
		TCγ' (CF before CD)	2×10^{-6}/yr	1.5	2.0	1.0	0.5	0.6
		TW γ' (CF before CD)	3×10^{-6}/yr	50	2.0	40	0.1	0.3
		Sum	5×10^{-6}/yr					
NUREG/CR-6295	Peach Bottom	RPB1[(b)] (CF at VB)	1.2×10^{-6}/yr	11.5	4.3	3.5	0.11	0.1
		RPB2[(b)] (CF at VB)	1.0×10^{-6}/yr	7.3	4.3	2.5	0.11	0.1
		RPB6[(b)] (CF at VB)	3×10^{-8}/yr	11.5	4.3	3.5	0.44	0.4
		Sum	2.2×10^{-6}/yr					
	LaSalle	RLAS1[(b)] (CF before VB)	6.3×10^{-6}/yr	58	13.5	4.8	0.16	0.17
		RLAS2[(b)] (CF at VB)	6.2×10^{-6}/yr	3.8	7.3	2.5	0.15	0.03
		RLAS3[(b)] (CF at VB)	1.2×10^{-6}/yr	16.9	6.3	5.8	0.11	0.07
		RLAS4[(b)] (CF before VB)	2.4×10^{-6}/yr	23.7	1.8	0.5	0.18	0.12
		Sum	1.6×10^{-5}/yr					
	Grand Gulf	RGG1[(b)] (CF at VB)	8.4×10^{-7}/yr	3.6	4	2.6	0.23	0.11
		RGG3[(b)] (Late CF)	1.2×10^{-6}/yr	14	4	13	0.15	0.01
		Sum	2×10^{-6}/yr					

(a) For NUREG-0773, this represents the interval of time between the decision to take protective measures and the start of the release; for NUREG/CR-6295, this represents the time between when the water level reaches 2 feet above the bottom of the active fuel and the start of the release.
(b) These source terms have multiple plumes, which have been summed here for ease of comparison.
CF = containment failure.
CD = core damage.
VB = reactor vessel branch.

- Where direct comparisons can be made (i.e., for bypass sequences in PWRs and containment failures before vessel breach in BWRs) the release fractions from NUREG/CR-6295 are significantly lower than those from NUREG-0773.

- For several sequences in NUREG/CR-6295, the release fractions appear to be comparable to or slightly greater than those from NUREG-0773 (e.g., PWR sequence RSEQ1 and BWR sequence RPB6 which have a release magnitude comparable to the largest PWR release and BWR release from NUREG-0773, respectively. However, the release frequencies reported in NUREG/CR-6295 for these sequences are one to two orders of magnitude lower than those from NUREG-0773, resulting in a lower risk impact.

- The release times and the difference in time between core uncovery and release to the atmosphere are generally comparable between the two studies.

Based on the comparisons provided above, the expected impacts, i.e., the frequency-weighted consequences, from the airborne pathway using the updated source term information would be much lower than previously predicted.

E.3.3.2 Other Pathway Impacts

Since the comparison of the new source term information to that used in the 1996 GEIS environmental impact projection shows that the amount of release of radioactive material in a severe accident is estimated to be less than estimated in the 1996 GEIS, the environmental impacts from the other pathways (contamination of open bodies of water, groundwater contamination, and the resulting economic impacts from any pathway) will also be less than estimated in the 1996 GEIS.

E.3.3.3 Conclusion

More recent source term information indicates that the timing from dominant severe accident sequences, as quantified in NUREG/CR-6295 (NRC 1997b), is comparable to the analysis forming the basis of the 1996 GEIS. In most cases, the release frequencies and release fractions are significantly lower for the more recent estimate. Thus, the environmental impacts used as the basis for the 1996 GEIS (i.e., the frequency-weighted consequences) are higher than the impacts that would be estimated using the more recent source term information.

It is worth noting that a significant effort is ongoing to re-quantify realistic severe accident source terms under the State-of-the Art Reactor Consequence Analysis (SOARCA) Project. Preliminary results indicate that source term timing and magnitude may be significantly lower than quantified in previous studies (NRC 2008a). This information will be incorporated, as appropriate, in future revisions of this document.

E.3.4 Impact of Power Uprates

Power uprates are defined as the process of increasing the maximum power level at which a nuclear power plant may operate. Although power uprates have been approved by the NRC since 1977, the effects of power uprates since 1996 were not taken into account for the GEIS. Extended power uprates began to be approved in 1998. For BWRs, it became common for a power uprate to be between 10 and 20 percent, and for PWRs, up to 5 percent. The purpose of this section is to provide an assessment of the impacts of power uprates on severe accident scenarios and their environmental impacts.

The process of license amendments for power uprates requires licensees to evaluate the effects of the uprate on the safety of the plant. Design-basis accidents were analyzed to determine the change in possible dose, should an accident occur. Most commonly, loss of coolant accidents, control rod drop accidents and fuel handling accidents were assessed. Whole body and thyroid doses were determined for the exclusion area boundary, the outer edge of the low population zone, and the main control room. These values must meet 10 CFR Part 100 and 10 CFR Part 50, Appendix A, General Design Criterion (GDC) 19 dose limits. The effects of power uprates on CDF and large early release frequency (LERF) are also assessed.

E.3.4.1 Airborne Pathway Impacts

Power uprates require using fuel with a higher percentage of uranium-235 or additional fresh fuel in order to derive more energy from the operation of the reactor. This results in a larger radionuclide inventory (particularly short-lived isotopes, assuming no change in burnup limits) in the core, than the same core at a lower power level. The larger radionuclide inventory represents a larger source term for accidents and can result in higher doses to offsite populations in the event of a severe accident. Typically, short-lived isotopes are the main contributor to early fatalities. As stated in NUREG-1449 (NRC 1993), short-lived isotopes make up 80 percent of the dose following early release.

LERF represents the frequency of sequences that result in early fatalities. Thus, the impact of a power uprate on early fatalities can be gauged by considering the impact of the uprate on the LERF metric. To this end, Table E-14 presents the change in LERF calculated by each licensee who has been granted a power uprate of greater than 10 percent. As can be seen, the increase in LERF ranges from a minimal impact to an increase of 30 percent (with a mean of 10.5 percent). This change is judged to be small to moderate.

Table E-14. Changes in LERF for Extended Power Uprates >10 Percent

Plant	Percent Increase in Power	Percent Increase in Internal Event LERF
Brunswick 1, 2	15	4.5
Clinton	20	5.5
Dresden 2, 3	17	10
Duane Arnold	15.3	16
Ginna	16.8	19
Hope Creek	15	30
Quad Cities 1, 2	17.8	4
Susquehanna 1, 2	13	<1
Vermont Yankee	20	5
Mean	16.4	10.5

E.3.4.2 Other Pathway Impacts

As discussed in previous sections, the change in impacts due to other pathways is viewed to be bounded by the change in the airborne pathway, consistent with the results obtained in the 1996 GEIS.

E.3.4.3 Conclusion

Power uprates would result in a small to (in some cases) moderate increase in the environmental impacts from a postulated accident. However, taken in combination with the other information presented in this appendix, the increases would be bounded by the 95 percent UCB values in Tables 5.10 and 5.11 of the 1996 GEIS.

E.3.5 Impact of Higher Fuel Burnup

There has been continued movement toward higher fuel burnup, to allow for more efficient utilization of the fuel and longer operating cycles. An environmental assessment (EA) was published by the NRC in 1988 on the effects of increased peak burnup (to 60 GWd/MT, 5 percent by weight uranium-235). NUREG/CR-5009 (NRC 1988) is the basis for the EA. NUREG/CR-6703 (NRC 2001a) is a more current analysis using updated designs and data, and peak burnup to 75 GWd/MT.

The purpose of this section is to include the updated information from NUREG/CR-6703 into the GEIS to account for the effect of current and possible future increased fuel burnup on postulated accidents. Future peak burnups being considered are 62 GWd/MT for PWRs and 70 GWd/MT for BWRs.

E.3.5.1 Airborne Pathway Impacts

The environmental impacts of accidents where high burnup fuel is being used (assuming no change in plant power level) are due to the effects of an increased inventory of long-lived fission products. Long-lived fission products contribute primarily to latent health effects, and thus latent fatalities are used here as a measure of the impact of higher burnup fuel. Since latent fatalities are directly scalable to dose, the assessment is based upon the increase in population dose due to the use of high burnup fuel.

NUREG/CR-6703 (NRC 2001a) analyzed design-basis accidents from full power for PWR and BWR reactors at different levels of fuel burnup. A PWR steam generator tube rupture and a BWR main steam line break were analyzed. Burnup was analyzed to 75 GWd/MT, at which point, fuel with more than 5 percent by weight uranium-235 would be required. As described on page 25 of that document, the models used do not account for natural processes and engineered safety features, so "more attention should be paid to trends in doses than to absolute values."

Table E-15 shows doses at the exclusion area boundary (EAB) and the total population dose stated in NUREG/CR-6703. The EAB dose includes contributions from inhalation, and external dose. The total population dose also includes contributions from contaminated foods as well. The increase in population dose is moderate (~38 percent) from 42 to 75 GWd/MT for PWRs. For BWRs, the net increase in population dose is small (~8 percent). Although the analysis in NUREG/CR-6703 is for design-basis accidents, the percentage increase in impacts would be generally similar for severe accidents.

Table E-15. LOCA Consequences as a Function of Fuel Burnup

Reactor Type	Peak-Rod Burnup (GWd/MT)	Individual Dose at 0.8 km[(a)] (rem)[(b)]	Mean Total Population Dose (person-rem)[(b)]
PWR	42	10	940,000
	50	10	1,100,000
	60	10	1,200,000
	62	10	1,200,000
	65	11	1,200,000
	70	11	1,300,000
	75	11	1,300,000
BWR	60	10	1,300,000
	62	10	1,300,000
	65	10	1,300,000
	70	11	1,400,000
	75	11	1,400,000

(a) 0.8 km = 0.5 mi.
(b) Note that these doses are on a per event basis, not a frequency (per year) basis.

E.3.5.2 Other Pathway Impacts

As discussed in previous sections, the change in impacts due to other pathways is viewed to be bounded by the change in the airborne pathway, consistent with the results obtained in the 1996 GEIS.

E.3.5.3 Conclusion

Increased peak fuel burnup from 42 to 75 GWd/MT for PWRs, and 60 to 75 GWd/MT for BWRs, results in small to moderate increases (up to 38 percent) in the environmental impacts in the event of a severe accident. However, taken in combination with the other information presented in this appendix, the increases would be bounded by the 95 percent UCB values in Tables 5.10 and 5.11 of the 1996 GEIS.

E.3.6 Impact from Accidents at Low Power and Shutdown Conditions

The 1996 GEIS did not include an assessment of the environmental impacts of accidents initiated at low power or shutdown conditions. These conditions include power levels less than 5 percent, shutdown (with or without maintenance or plant modifications under way), and fuel handling. The safety concern under these conditions is that plant configurations may be established where not all plant safety systems and features would be operable (e.g., containment integrity may not be required), and activities (e.g., plant modification) could be under way that could not be done while at full power. Accordingly, accidents initiated at such conditions may have different initiators, progress differently, and have different consequences than those initiated at full power conditions. In addition, operating experience has shown that events affecting fuel cooling do occur during shutdown operation. Accordingly, the industry implemented a number of voluntary measures in response to NRC generic letters and bulletins, and in 1991 developed guidelines for the assessment of shutdown management and implementation of safety improvements (NUMARC 1991). As discussed in SECY-97-168 (NRC 1997c), these voluntary industry initiatives resulted in improved safety.

The purpose of this section is to provide an assessment of the risk from postulated severe accidents at low power and shutdown conditions relative to the risk from postulated severe accidents at full power conditions, including a comparison against the findings in the 1996 GEIS.

The conditions assessed are:

- Plant operation at power levels between 0 and 5 percent;
- Shutdown with containment open; and
- Fuel handling inside the containment structure.

Several sources of information are available to support this assessment. These include studies that have been done assessing actual events and the risk from accidents at low power and shutdown conditions. These studies are: (1) NUREG-1449 (NRC 1993); (2) NUREG/CR-6143 (NRC 1995b); and (3) NUREG/CR-6144 (NRC 1995a). In addition, in 1997, the NRC staff recommended a proposed rule be considered to address shutdown conditions. Although the Commission did not approve going forward with the proposed rule (see SRM-97-168, NRC 1997d), the technical basis for the proposed rule provides additional useful information.

E.3.6.1 Airborne Pathway Impacts

NUREG-1449 (NRC 1993) presents an analysis of actual events that have occurred at low power and shutdown conditions. This analysis includes an estimate of the conditional core damage frequency associated with each event and an overall assessment of the range of total core damage frequencies (mean value) that could result from events at low power and shutdown conditions. This range was from 10^{-5}/yr to 10^{-4}/yr.

NUREG/CR-6143 (NRC 1995b) and NUREG/CR-6144 (NRC 1995a) provide low power and shutdown risk assessments for two plants (Grand Gulf and Surry). For Grand Gulf, the mean core damage frequency stated in NUREG/CR-6143 is approximately 2×10^{-6}/yr and for Surry (NUREG/CR-6144) it is 4×10^{-6}/yr. However, such core damage frequencies need to be considered with respect to their consequences. Due to the decay time associated with low power and shutdown conditions (i.e., decay of short-lived isotopes and lower decay heat) and, in most cases, longer times available to take mitigative action, the offsite consequences would be less than for accidents from full power. However, in certain plant operating states, the containment in those states may be open. Thus, a higher conditional probability for containment bypass might exist.

NUREG/CR-6143 and NUREG/CR-6144 also provide estimates of the offsite airborne pathway consequences on human health from accidents (internal events only) at low power and shutdown conditions. Tables E-16 and E-17 list these estimates for Grand Gulf and Surry, respectively. Also shown for each plant are the airborne pathway offsite consequence results for accidents from full power from NUREG-1150 (NRC 1990b) (for internal events) and from the 1996 GEIS. As can be seen, the airborne impacts (airborne pathway risk and probability-weighted consequences) from accidents at low power and shutdown are comparable to those from full power, as quantified in these studies. Although the impacts for low power and shutdown conditions are somewhat greater (by about a factor of 2 to 5) for certain metrics, these differences are small in an absolute sense. Moreover, the airborne impacts of accidents from low power and shutdown are significantly less than those stated in the 1996 GEIS (by more than an order of magnitude). Thus, even though the 1996 GEIS estimates regarding the airborne pathway environmental impact are for internal events at full power only, their conservatism causes them to bound the impacts from accidents at low power and shutdown.

E.3.6.2 Other Pathway Impacts

For the impacts from surface water and groundwater contamination from accidents at low power and shutdown, the estimates for accidents from full power (internal events only) in the 1996 GEIS can be used for comparison. In the 1996 GEIS, for the surface water pathways, it was estimated that the impacts from the drinking water pathway would be a small fraction of those for the airborne pathway. The risk associated with the aquatic food pathway was found to be

Table E-16. Airborne Impacts of Low Power and Shutdown Accidents (Grand Gulf)

Impact	Low Power/Shutdown Accidents NUREG/CR-6143 (95th percentile values)	Full Power Accidents Internal Events NUREG-1150 (95th percentile values)	Full Power Accidents Internal Events 1996 GEIS (95th percentile values)
Individual risk			
EF[(a)] (1 mi)	~3×10^{-10}/yr	~1.5×10^{-10}/yr	
LF[(b)] (10 mi)	~5×10^{-9}/yr	~1×10^{-9}/yr	
Total person-rem per year (entire region)	~28	~15	1,441
Total early fatality risk	~4×10^{-8}/yr	~2.5×10^{-8}/yr	2.8×10^{-3}/yr
Total latent fatality risk	~1×10^{-2}/yr	~2.5×10^{-3}/yr	1.0/yr
CDF	5.6×10^{-6}/yr	1.2×0^{-5}/yr	2.4×10^{-5}/yr[(c)]

(a) EF = early fatality risk. The individual early fatality risk within one mile (1.6 km) is the frequency (per year) that a person living within one mile (1.6 km) of the site boundary will die within a year due to the accident. The entire population within one mile is considered to obtain an average value.
(b) LF = latent fatality risk. The individual latent cancer fatality risk within 10 miles (16 km) is the frequency (per year) that a person living within 10 miles (16 km) of the plant will die many years later from cancer due to radiation exposure received from the accident. The entire population within 10 miles (16 km) is considered to obtain an average value.
(c) This is the CDF from the Grand Gulf original EIS.

Table E-17. Airborne Impacts of Low Power and Shutdown Accidents (Surry)

Impact	Low Power/Shutdown Accidents (NUREG/CR-6144) (95th percentile values)	Full Power Accidents Internal Events NUREG-1150 (95th percentile values)	Full Power Accidents Internal Events 1996 GEIS (95th percentile values)
Individual risk			
EF[(a)] (1 mi)	~7×10^{-9}/yr	~4×10^{-8}/yr	
LF[(b)] (10 mi)	~7×10^{-9}/yr	~1×10^{-8}/yr	
Total person-rem per year (entire region)	~1.3	~150	1,200
Total early fatality risk	~2×10^{-7}/yr	~4×10^{-6}/yr	1.6×10^{-2}/yr
Total latent fatality risk	~5×10^{-2}/yr	~2.5×10^{-2}/yr	0.9/yr
CDF	1.9×10^{-5}/yr	1.3×10^{-4}/yr	

(a) EF = early fatality risk. The individual early fatality risk within one mile (1.6 km) is the frequency (per year) that a person living within one mile (1.6 km) of the site boundary will die within a year due to the accident. The entire population within one mile is considered to obtain an average value.
(b) LF = latent fatality risk. The individual latent cancer fatality risk within 10 miles (16 km) is the frequency (per year) that a person living within 10 miles (16 km) of the plant will die many years later from cancer due to radiation exposure received from the accident. The entire population within 10 miles (16 km) is considered to obtain an average value.

also relatively small compared to the risks associated with the airborne pathway for most sites and essentially the same as the atmospheric pathway for the few sites with large annual aquatic food harvests. With the airborne impacts from accidents at low power and shutdown in NUREG/CR-6143, -6144, and NUREG-1150 estimated to be considerably less than the impacts from accidents at full power in the 1996 GEIS, the surface water pathway impacts should likewise be less, and thus, the risks portrayed in the 1996 GEIS should be bounding.

Section 5.3.3.4 of the 1996 GEIS concluded that the contribution of risk from the groundwater pathway for at-power accidents "generally contributes only a small fraction of that risk attributable to the atmospheric pathway but in a few cases may contribute a comparable risk." Groundwater contamination due to basemat melt-through would be less likely than for accidents at full power, due to the lower decay heat associated with low power and shutdown events. Thus, the risks portrayed in the 1996 GEIS are considered to be bounding.

With respect to the economic impacts regardless of contamination pathway, the lower estimated person-rem/yr from accidents at low power and shutdown should also result in lower economic impacts than from accidents at full power.

E.3.6.3 Conclusion

In summary, it is concluded that the environmental impacts from accidents at low power and shutdown conditions are generally comparable to those from accidents at full power when comparing the NUREG/CR-6143 (NRC 1995b) and NUREG/CR-6144 (NRC 1995a) values to NUREG-1150 (NRC 1990b) values. Although the impacts for low power and shutdown conditions could be somewhat greater than for full power (for certain metrics), the 1996 GEIS estimates of the environmental impact of severe accidents bound the potential impacts from accidents at low power and shutdown with margin. Finally, as cited above and discussed in SECY-97-168 (NRC 1997c), industry initiatives taken during the early 1990s have also contributed to the improved safety of low power and shutdown operation.

E.3.7 Impact from Accidents at Spent Fuel Pools

The 1996 GEIS did not include an explicit assessment of the environmental impacts of accidents at the spent fuel pools (SFPs) located at each reactor site. The 1996 GEIS did, however, discuss qualitatively (see Section 5.2.3.1) the reasons why the impact of accidents at SFPs would be much less than that from reactor accidents. Thus, in Table B-1 of 10 CFR Part 51, it was concluded that accidents at SFPs could be classified as Category 1 and not require further analysis in support of license renewal. This was primarily due to the fact that the resolution of Generic Safety Issue 82, "Beyond Design Basis Accidents in Spent Fuel Pools," concluded that the risk from accidents at SFPs was low and, accordingly, no additional regulatory action was necessary. The analysis supporting this conclusion is contained in NUREG-1353 (NRC 1989).

Since issuance of the 1996 GEIS, additional analysis of the risk from spent fuel pool accidents has been performed and documented. For example, in 2001, the NRC published NUREG-1738 (NRC 2001b), which evaluated SFP risk during decommissioning. As a result of the September 11, 2001, terrorist attacks, additional analysis has been performed on spent fuel pool (SFP) security, although much of this work is security-related information and not publically available. In addition, there are two other major activities of note: (1) a 2004 to 2005 study performed by the National Academies (National Research Council 2006b), and (2) a 2006 Petition for Rulemaking (see NRC 2008d).

The purpose of this section is to consider the risk from severe accidents in SFPs relative to the risk from severe accidents in reactors, including a comparison against the findings in the 1996 GEIS. The impacts considered are only those from spent fuel in the pool. Spent fuel assembly dry cask safety is not included, since cask safety is addressed under 10 CFR Part 72.

E.3.7.1 Airborne Pathway Impacts

The analysis contained in NUREG-1738 (NRC 2001b) assesses the impacts from accidents at a typical SFP at decommissioning nuclear power plants. The impacts assessed are those associated with the airborne pathway impact on human health. The analysis covers a range of decay times for the fuel stored in the pool, a number of initiating events, and some variations in emergency evacuation times, fission product releases, and seismic hazard. The initiating events included in the analysis are listed below.

- Seismic (for central and eastern U.S. sites)[k]
- Cask drop
- Loss of offsite power
- Internal fire

(k) The seismic risk analysis performed in NUREG-1738 was based on site-specific seismic hazard estimates for nuclear power plants in the central and eastern United States found in NUREG-1488, "Revised Livermore Seismic Hazard Estimates for 69 Nuclear Power Plant Sites East of the Rocky Mountains." As such, nuclear power plants in the western United States, such as Diablo Canyon, San Onofre, and Columbia, were not specifically considered in this study. Nothing in NUREG-1738, or the staff's reliance on it here, undermines the staff's initial conclusion in the 1996 GEIS that the impacts of SFP severe accidents will be comparable to reactor severe accidents for all facilities.

- Loss of pool cooling
- Loss of pool coolant inventory
- Accidental aircraft impact (although not deliberate impacts)
- Tornado missile

The SFP inventory assumed was 3½ core loads with an average fuel burnup of 60 GWd/MT. Although intended to be representative of the SFP in a typical decommissioning PWR or BWR, the assumed core inventory, burnup, and decay time range is also reasonably representative of that for operating PWRs and BWRs while at power. In addition to the above results, NUREG-1738 also assessed the risk from recriticality in the SFP and concluded that, given licensee surveillance and monitoring programs, the potential risk of such events is small.

The analysis conducted in NUREG-1738 assumed the plant was in its decommissioning phase and, thus, has fewer protective features for the prevention or mitigation of SFP accidents. Therefore, the impact analysis contained in NUREG-1738 is considered conservative. In addition, the NUREG-1738 impact analysis assumed that the zirconium fuel cladding would start to burn and the event would be nonrecoverable when the water level in the pool falls to within 3 feet (1 m) above the top of the assemblies' active fuel region. This is also conservative and does not credit potential operator actions to prevent or mitigate SFP accidents beyond that point, or the fact that for a wide range of conditions spent fuel can be air-cooled. Table E-18 summarizes the airborne pathway impact on human health from a severe accident in a SFP (from the NUREG-1738 analysis) for a time period of 1 month to 2 years (i.e., a typical operating reactor fuel cycle). Ranges are given to account for differences in emergency planning and seismic hazard assumptions. The site characteristics used in NUREG-1738 were those from the Surry plant. Thus Table E-18 also presents Surry's site-specific results from NUREG-1150 (NRC 1990b) and the 1996 GEIS.

As can be seen in Table E-18, the impacts from SFP accidents at Surry (as calculated in NUREG-1738) are generally comparable to or smaller than the analogous NUREG-1150 internal event reactor accidents when using the low ruthenium release source term.[(l)] For the high ruthenium release source term, the NUREG-1738 results are generally higher than the accompanying reactor results from NUREG-1150. For either source term, the NUREG-1738

(l) Due to a concern about the potential release of ruthenium isotopes from the spent fuel stored in the SFP, two sensitivity cases were analyzed in NUREG-1738: one with a ruthenium release fraction of 2×10^{-5} (called the base case or the low ruthenium release case) and another with a ruthenium release fraction of 1.0 (called the high ruthenium release case).

Table E-18. Impacts of Accidents at SFPs from NUREG-1738[a]

	Spent Fuel Pools[b] (1 month to 2 years decay time)		Reactors		
	NUREG-1738 Low Ru Release (range of means)	**NUREG-1738 High Ru Release (range of means)**	**NUREG-1150 Surry (mean)**	**NUREG-1150 Surry (95th percentile)**	**1996 GEIS Surry (95th percentile)**
Individual risk					
EF[c] (1 mi)	2×10^{-9} to 7×10^{-9}/yr	6×10^{-8} to 1×10^{-7}/yr	1.5×10^{-8}/yr	4×10^{-8}/yr	
LF[d] (10 mi)	1×10^{-8}/yr	2×10^{-7}/yr	1.5×10^{-9}/yr	1×10^{-8}/yr	
Total person-rem per year	2.5 to 12 (50 mi)	8 to 60 (50 mi)	6 (50 mi) 30 (entire region)	30 (50 mi) 150 (150 mi)	1,200 (150 mi)
Total early fatality risk	2×10^{-7} to 6×10^{6}/yr	1×10^{-5} to 5×10^{-4}/yr	1×10^{-6}/yr	3×10^{-6}/yr	1.6×10^{-2}/yr

(a) All values are approximate.
(b) Values are obtained from Figures 3.7-3, 3.7-4, 3.7-7, and 3.7-8 of NUREG-1738.
(c) EF = early fatality risk. The individual early fatality risk within one mile (1.6 km) is the frequency (per year) that a person living within one mile (1.6 km) of the site boundary will die within a year due to the accident. The entire population within one mile (1.6 km) is considered to obtain an average value.
(d) LF = latent fatality risk. The individual latent cancer fatality risk within 10 miles (16 km) is the frequency (per year) that a person living within 10 miles (16 km) of the plant will die many years later from cancer due to radiation exposure received from the accident. The entire population within 10 miles (16 km) is considered to obtain an average value.

impacts are much less than the conservative estimates of full power reactor accidents at Surry as estimated in the 1996 GEIS.

The impacts stated in NUREG-1738 are also similar to those calculated for the resolution of Generic Safety Issue 82, in which NUREG-1353 (NRC 1989) calculated a best-estimate population dose of 16 person-rem per year.[m] While the NUREG-1738 results are for the Surry site, individual risk metrics for early fatalities and latent fatalities should be relatively insensitive to the site-specific population (see pg. 3-28 of NUREG-1738) because these metrics reflect doses to the close-in population. In addition, while results are presented for both the low and high ruthenium source term, the low ruthenium source term is still viewed as the more accurate representation. Therefore, the risk and environmental impact from fires in SFPs as analyzed in NUREG-1738 are expected to be comparable to or lower than those from reactor accidents and are bounded by the 1996 GEIS.

Since the issuance of NUREG-1738 (NRC 2001b), and subsequent to the terrorist attacks of September 11, 2001, significant additional analyses have been performed that support the view that the risk of a successful terrorist attack (i.e., one that results in a zirconium fire) is very low at all plants. These analyses were conducted by the Sandia National Laboratories and are

(m) Taken from the Executive Summary of that report: total dose = 8×10^{6} person-rem; event frequency = 2×10^{-6} per year.

collectively referred to herein as the "Sandia studies." The Sandia studies are sensitive, security-related information and are not available to the public. The Sandia studies considered spent fuel loading patterns and other aspects of a pressurized-water reactor SFP and a boiling-water reactor SFP, including the role that the circulation of air plays in the cooling of spent fuel. The Sandia studies indicated that there may be a significant amount of time between the initiating event (i.e., the event that causes the SFP water level to drop) and the spent fuel assemblies becoming partially or completely uncovered. In addition, the Sandia studies indicated that for conditions where air cooling may not be effective in preventing a zirconium fire, there is a significant amount of time between the spent fuel becoming uncovered and the possible onset of such a zirconium fire, thereby providing a substantial opportunity for both operator and system event mitigation.

The Sandia studies, which more fully account for relevant heat transfer and fluid flow mechanisms, also indicated that air cooling of spent fuel would be sufficient to prevent SFP zirconium fires at a point much earlier following fuel offload from the reactor than previously considered (e.g., in NUREG-1738). Thus, the fuel is more easily cooled, and the likelihood of a zirconium fire is therefore reduced.

Furthermore, additional mitigation strategies implemented subsequent to September 11, 2001, enhance spent fuel coolability and the potential to recover SFP water level and cooling prior to a potential zirconium fire. The Sandia studies also confirmed the effectiveness of these additional mitigation strategies to maintain spent fuel cooling in the event the pool is drained and its initial water inventory is reduced or lost entirely. Based on the more rigorous accident progression analyses, the recent mitigation enhancements, and NRC site evaluations of every SFP in the United States, the risk of an SFP zirconium fire initiation is expected to be less than reported in NUREG-1738 (NRC 2001b) and previous studies. For additional information on SFP safety and security, the reader is referred to the NRC's response to a National Academy of Sciences study on the topic (NRC 2005a) and the NRC's response to a petition for rulemaking (NRC 2008d).

E.3.7.2 Other Pathway Impacts

The NUREG-1738 (NRC 2001b) analysis did not address the impacts with respect to the other pathways (open bodies of water and groundwater). The 1996 GEIS estimated these impacts for reactor accidents from full power (internal events only) using the results from plant-specific reactor accident analysis to assess contamination of open bodies of water and from the Liquid Pathway Generic Study (NUREG-0440; NRC 1978) to assess the contamination of groundwater from basemat melt-through accidents.

In both cases, the impacts on human health from surface water and groundwater contamination are only a small fraction of those impacts from the airborne pathway, except in a few cases where the impacts are comparable. With the impacts from the airborne pathway associated with spent fuel pool accidents (as stated in NUREG-1738) being comparable to the impacts from reactor accidents, as stated in NUREG-1150 (NRC 1990b), the impacts from SFP-related surface water and groundwater contamination may also be comparable, even though the SFP fuel inventory is several times that of the reactor. This is due to the lower probability of occurrence of SFP accidents, the effects of decay of the fission products on the radionuclide inventory, and the lower energy density of the fuel inventory, which makes basemat melt-through more unlikely.

The same conclusion can also be drawn with respect to the economic impacts. These impacts are related to the likelihood of the accidents and the cost of cleanup and food interdiction. Even with higher fuel inventories, the lower likelihood of accidents in the SFP reduces the economic impacts. For example, the UCB economic impact identified in Table 5.31 in the 1996 GEIS from full power reactor accidents at Surry is approximately $1.1 million/yr. The worst-case economic impacts estimated in past studies for SFP accidents ranged from approximately $18,000/yr to $120,000/yr.[(n)]

An issue related to the groundwater pathway that has received significant attention since the issuance of the 1996 GEIS is leakage of water from SFPs (or related systems) at Salem Unit 1, Indian Point Units 1 and 2, and Seabrook. Instances of this kind are adequately monitored and addressed via existing regulatory programs, and do not fall within the scope of this section. For more information on this topic, the reader is referred to NUREG-0933, Supplement 31, Section 3, Issue 202 (NRC 2007c) and NRC 2008b.

E.3.7.3 Conclusion

In summary, it is concluded that the environmental impacts from accidents at SFPs (as quantified in NUREG-1738 [NRC 2001b]) can be comparable to those from reactor accidents at full power (as estimated in NUREG-1150 [NRC 1990b]). Subsequent analyses performed, and mitigative measures employed since 2001, have further lowered the risk of this class of accidents. In addition, even the conservative estimates from NUREG-1738 are much less than the impacts from full power reactor accidents as estimated in the 1996 GEIS. Therefore, the environmental impacts stated in the 1996 GEIS bound the impact from SFP accidents.

E.3.8 Impact of the Use of BEIR VII Risk Coefficients

Section 5.3.3.2.2 from the 1996 GEIS discussed adverse health effects from exposure to radiation and referenced several National Academy of Sciences reports (BEIR I, III, and V)

(n) The former estimate uses information from Tables C.95 and C.101 of NUREG/BR-0184 (NRC 1997a), while the latter uses information from Tables 5.1.1 and 5.1.2 of NUREG-1353 (NRC 1989).

(National Research Council 1972, 1980, 1990) as sources of risk coefficients for fatal cancers (i.e., latent fatalities) associated with radiation exposure. Benchmark evaluations of the exposure index methodology employed by the 1996 GEIS were conducted using the MELCOR Accident Consequence Code System (MACCS), as described in Section 5.3.3.2.3 of the original GEIS. MACCS is the predecessor of the currently used MACCS2 code, and represented the state-of-the-art for assessing risks associated with postulated severe reactor accidents at the time of the original GEIS. That study used a linear cancer model based on the BEIR V report (National Research Council 1990). The code-to-code comparisons suggest that latent fatality values in the FESs are an order of magnitude too low. Therefore, to account for this, the latent fatality results predicted from the FES values were multiplied by a factor of 10 to obtain the final predicted latent fatality results in the 1996 GEIS. This adjustment in combination with the use of 95th percentile UCB values ensured that the basis for health effects would be conservative.

In 2006, the National Research Council's Committee on the Biological Effects of Ionizing Radiation (BEIR) published BEIR VII, entitled *Health Risks from Exposure to Low Levels of Ionizing Radiation* (National Research Council 2006a). BEIR VII provides estimates of the risk of incidence and mortality for males and females (see Section 3.9.1.4 and Appendix D of this report for more information). The BEIR VII report estimates that the fatal cancer risk coefficient is approximately 20 percent higher than the International Commission on Radiological Protection (ICRP) recommendation (as described in ICRP 1991). The difference of 20 percent is within the margin of uncertainty associated with these estimates (see Appendix D.8.1.4 for a detailed discussion of the BEIR VII report).

The NRC staff completed a review of the BEIR VII report and documented its findings in NRC 2005b. In this paper, the NRC staff concluded that the findings presented in the BEIR VII report agree with the NRC's current understanding of the health risks from exposure to ionizing radiation. The NRC staff agreed with the BEIR VII report's major conclusion that current scientific evidence is consistent with the hypothesis that there is a linear, no-threshold dose response relationship between exposure to ionizing radiation and the development of cancer in humans. This conclusion is consistent with the process the NRC uses to develop its standards of radiological protection. Therefore, the NRC's regulations continue to be adequately protective of public health and safety and the environment. This general topic is discussed further in a 2007 denial of a Petition for Rulemaking, as discussed in NRC 2007d.

E.3.9 Uncertainties

Section 5.3.5 in the 1996 GEIS provides a discussion of the uncertainties associated with the analysis in the GEIS and in the individual plant EISs used to estimate the environmental impacts of severe accidents. The uncertainties discussed covered:

- The probability of an accident.

- The quantity and chemical form of radioactivity released.

- Atmospheric dispersion modeling for the radioactive plume transport, including:
  - duration, energy release, and in-plant radionuclide decay time;
  - meteorological sampling scheme used;
  - emergency response effectiveness and warning time;
  - dose conversion factors and dose-response relationships for early health consequences;
  - dose conversion factors and dose-response relationships for latent health consequences;
  - chronic exposure pathways; and
  - economic data and modeling.

- Assumption of normality for random error components.

- The exposure-index method, and
  - selection of exposure index parameters;
  - selection of distances;
  - regressing early fatalities for only large plants; and
  - normalization of plants for latent fatalities, costs, and dose.

The 1996 GEIS recognized that the uncertainties in the estimated impacts could be large (i.e., from a factor of 10 to 1000). Reference was made to NUREG-1150 (NRC 1990b) as providing more state-of-the-art risk analysis that also considered uncertainties and that the cumulative effect of this analysis shows a reduction in risk.

In an attempt to help compensate for uncertainties, the 1996 GEIS used very conservative estimates of environmental impacts. These included:

- Use of the 95th percentile confidence values in estimating airborne pathway and economic impacts;

- Use of site-specific analysis for estimating surface water pathway impacts; and

- Use of NUREG-0440 (NRC 1978) results to bound the estimated groundwater pathway impacts.

It was generally concluded that even with uncertainties, the environmental impacts estimated in the 1996 GEIS were adequate for use.

Many of these same uncertainties also apply to the analysis used in this update. However, as discussed in Sections E.3.1 through E.3.8 of this revision, more recent information is used to supplement the estimate of the environmental impacts contained in the 1996 GEIS. In effect, the assessments contained in Sections E.3.1 through E.3.8 of this revision provide additional information and insights into items that could be considered areas of uncertainty associated with the 1996 GEIS.

This more recent information also provides insights on additional sources of uncertainty from those discussed in the 1996 GEIS. Each of these insights on additional sources of uncertainty is discussed below.

E.3.9.1 Emergency Planning (EP)

The 1996 GEIS (in Section 5.3.5.3) included a discussion on uncertainties associated with EP. However, no quantitative information on the magnitude of these uncertainties was presented. To provide a perspective on the magnitude of the uncertainty, the following information is provided.

NUREG-1150 (NRC 1990b) and the SFP accident analysis in NUREG-1738 (NRC 2001b) specifically assessed the effect of different EP assumptions on the airborne pathway impacts. NUREG-1150 assessed four alternative emergency response modes in addition to its base case (99.5 percent of the population within 10 mi was evacuated in 4.5 hours with no sheltering). These alternatives were assessed for reactor accidents from full power, with the Surry and Peach Bottom analyses including seismic and fire initiated events as well as internal events. For the worst case (no evacuation, no sheltering, and early relocation), the estimated early fatalities per year were approximately a factor of 10 higher than the base case.

The SFP accident analysis in NUREG-1738 also specifically assessed the effect of variations in emergency evacuation. The variations were assessed against the base case used in the NUREG-1150 risk analysis. Doses beyond 20 mi were not calculated. Cases where the evacuation was faster, slower, and where fewer people were evacuated were assessed. As can be expected, improved evacuation scenarios resulted in smaller impacts, and relaxed evacuation scenarios resulted in additional impacts. The impacts associated with relaxed evacuation scenarios did go up, but only a few percent in societal dose (i.e., person-rem) and up to a factor of 10 in early fatalities. However, these impacts are still far below the conservative characterization of the impacts for reactor accidents contained in the original GEIS.

E.3.9.2 Population Increase

The assessments of environmental impacts contained in NUREG-1150 (NRC 1990b), NUREG-1738 (NRC 2001b), NUREG-1449 (NRC 1993), NUREG/CR-5305 (NRC 1992), NUREG/CR-6143 (NRC 1995b) and NUREG/CR-6144 (NRC 1995a) are all based on populations that existed in the mid-1980s to mid-1990s. The 1996 GEIS estimated impacts at the mid-year of each plant's license renewal period (i.e., 2030 to 2050). To adjust the impacts estimated in the NUREGs and NUREG/CRs to the mid-year of the assessed plant's license renewal period, the information (i.e., exposure indexes [EIs]) in the 1996 GEIS can be used. The EIs adjust a plant's airborne and economic impacts from the year 2000 to its mid-year license renewal period based on population increases. These adjustments result in anywhere from a 5 to a 30 percent increase in impacts, depending upon the plant being assessed. Given the range of uncertainty in these types of analyses, a 5 to 30 percent change is not considered significant. Therefore, the effect of increased population around the plant does not generally result in significant increases in impacts.

E.4 Severe Accident Mitigation Alternatives (SAMAs)

In Section 5.4 of the 1996 GEIS, the purpose and role of severe accident mitigation design alternatives (SAMDAs) in the license renewal process are discussed. Severe accident mitigation alternatives (SAMAs) include design alternatives (SAMDAs) and alternatives that involve changes in procedures and training. With respect to this revision of the GEIS, the purpose and objectives of SAMAs remain unchanged.

The purpose of this section is to discuss the impacts on SAMA analyses of the assessments presented in this revision. It should be noted that since publication of this 1996 GEIS, many improvements have occurred that have enhanced reactor safety. These are discussed in Section E.2 of this revision and, as can be seen in improved plant performance measures, have been effective. Even so, the SAMA analyses that have been performed to date have found SAMAs that were cost-beneficial, or at least possibly cost-beneficial subject to further analysis, in approximately half of the plants. However, none of the SAMAs identified related to managing the effects of aging during the period of extended operation. Therefore, they did not need to be implemented as part of license renewal, pursuant to the regulations in 10 CFR Part 54. In general, the cost-beneficial SAMAs were identified for further evaluation by the licensee under the current operating license. In several cases, the applicant has decided to implement the modifications even though they were not related to license renewal (NRC 2006a).

The SAMA analysis performed in support of license renewal has focused on those areas of greatest risk (accidents initiated by internal and external events) and on measures that could result in the greatest risk reduction in a cost-beneficial fashion. Even though the 1996 GEIS did not explicitly consider accidents initiated by external events in estimating the environmental impacts from severe accidents, the environmental impacts from external events are included in an applicant's SAMA analysis for license renewal by following the guidance contained in NEI 05-01, Revision A (NEI 2005). This guidance (which is endorsed by the NRC in Regulatory Guide 4.2, Supplement 1, Revision 1, "Preparation of Environmental Reports for Nuclear Power Plant License Renewal Applications," [NRC 2013]) calls for the consideration of external events in assessing SAMAs. External events are considered by multiplying the internal event risk by a factor that accounts for any increase in risk caused by external events. The multiplication factor is determined on a plant-specific basis considering previous and current external event analyses (e.g., IPEEE). Given the existing information on the contribution to risk from external events, the approach described in NEI 05-01 continues to be a reasonable approach to address the external event risk contribution.

This GEIS revision has assessed other potential contributors to risk. Therefore, it is reasonable to assess whether those contributors should be included in the SAMA analysis. Specifically, these contributors are:

- Power uprates;
- The use of higher burnup fuel;
- Accidents from low power and shutdown conditions; and
- Accidents at SFPs.

With respect to power uprates and the use of higher burnup fuel, the increased impacts are small compared to the impacts in the 1996 GEIS, and these factors are included in any severe accident assessment for license renewal. Therefore, no additional SAMA analysis is required.

With respect to accidents from low power and shutdown conditions (which are not currently included in SAMA analysis), the CDFs are generally lower and the risks are comparable to those of accidents from full power. In addition, there have been industry initiatives to improve low power and shutdown safety. It is also likely that some SAMAs identified as a result of assessing risks from accidents at full power would provide benefits to accidents from low power. Therefore, the potential for cost-beneficial SAMAs related to low power and shutdown accidents is considered to be less than for accidents at full power. Accordingly, it is reasonable to continue to exclude low power and shutdown conditions from SAMA analysis consideration.

With respect to accidents in SFPs, the additional mitigative measures implemented following the attacks of September 11, 2001, have further lowered the risk of this class of accidents, and therefore make the potential for finding cost-effective SAMAs related to SFP accidents substantially less than for reactor accidents. Therefore, it is reasonable to conclude that accidents at SFPs do not need to be considered in the SAMA analysis.

With respect to which plants must submit a SAMA analysis, 10 CFR 51.53(c)(3)(ii)(L) states that, "[i]f the staff has not previously considered severe accident mitigation alternatives for the applicant's plant, in an environmental impact statement or related supplement or in an environmental assessment, a consideration of alternatives to mitigate severe accidents must be provided." Applicants for plants that have already had a SAMA analysis considered by the NRC as part of an EIS, supplement to an EIS, or EA, do not need to have a SAMA analysis reconsidered for license renewal. In forming its basis for determining which plants needed to submit a SAMA, the Commission noted that all licensees had undergone, or were in the process of undergoing, more detailed site-specific severe accident mitigation analyses through processes separate from license renewal, specifically the Containment Performance Improvement (CPI), Individual Plant Examination (IPE), and IPE for external events (IPEEE) programs (61 FR 28467). In light of these studies, the Commission stated that it did not expect future SAMA analyses to uncover "major plant design changes or modifications that will prove to be cost-beneficial." (61 FR 28467). The NRC's experience in completed license renewal proceedings has confirmed this prediction. As a result, the totality of these studies (the former SAMA analyses, the IPE, the IPEEE, and the CPI) provides a strong basis for the Commission's decision to not require applicants to perform an additional SAMA analysis in a license renewal application if the NRC had previously evaluated one for that plant. Therefore, applicants for license renewal of those plants that have already had a SAMA analysis considered by the NRC as part of an EIS, supplemental to an EIS or EA, need not perform an additional SAMA analysis for license renewal.

E.5 Summary and Conclusion

The 1996 GEIS estimated the environmental impacts on human health and economic factors from full power severe reactor accidents initiated by internal events. Sections E.3.1 through E.3.8 of this revision assessed the impacts of new information and additional accident considerations on the environmental impact of severe accidents contained in the 1996 GEIS. In addition, the impact of uncertainties associated with the new information is assessed in Section E.3.9. The purpose of this section is to discuss the aggregate effect of the new information on the environmental impacts and uncertainties stated in the 1996 GEIS and to state what conclusions can be drawn.

The different sources of new information can be generally categorized by their effect of decreasing, not affecting, or increasing the best-estimate environmental impacts associated with postulated severe accidents. Those areas where a decrease in best-estimate impacts would be expected are:

- New internal events information (decreases by an order of magnitude)

- New source term information (significant decreases)

Areas likely leading to either a small change or no change include:

- Use of BEIR VII risk coefficients

Lastly, those areas leading to an increase in best-estimate impacts would consist of:

- Consideration of external events (comparable to internal event impacts)

- Power uprates (small to moderate increase)

- Higher fuel burnup (small to moderate increases)

- Low power and shutdown events (could be comparable to full power event impacts)

- Spent fuel pool accidents (could be comparable to full power event impacts)

Given the difficulty in conducting a rigorous aggregation of these results (due to the differences in the information sources utilized), a fairly simple approach is taken. The latter group contains three areas where the increase could be comparable to the current risk and two areas where the increase could approach 30 to 40 percent. The net increase from these five areas would therefore be approximately 470 percent[o] (increase by a factor of 4.7). The reduction in risk due to newer internal event information would account for a decrease by a factor of 5 to 100. The net effect of an increase on the order of 500 percent and a decrease on the order of 500 percent to 10,000 percent would be a reduction in estimated impacts (as compared to the 1996 GEIS assessment).

Furthermore, even if one assumed that the net effect of the new information was no change in risk, the information provided throughout this appendix has demonstrated that the level of conservatism in the upper bound estimates utilized in the 1996 GEIS is much larger than the individual (or cumulative) deltas from the updated information. In particular, Section E.3.1 demonstrates that the GEIS values were a factor of 2 to 4 higher than the underlying EIS values.

(o) This approximation simply assumes that each comparable area results in an increase of 100 percent and the other two areas (uprates and burnup) each result in an increase of 35 percent.

With respect to uncertainties, the 1996 GEIS contained an assessment of uncertainties in the information used to estimate the environmental impacts. Section 5.3.5 of the 1996 GEIS discusses the uncertainties and concludes that they could cause the impacts to vary anywhere from a factor of 10 to a factor of 1,000. This range of uncertainties bounds the uncertainties discussed in Section E.3.9 above, which ranged from a factor of 3 to 10, as well as the uncertainties brought in by the other sources of new information.

Given the discussion in this appendix, the staff concludes that the reduction in environmental impacts from the use of new information (since the 1996 GEIS analysis) outweighs any increases resulting from this same information. As a result, the findings in the 1996 GEIS remain valid. Therefore, design-basis accidents remain a Category 1 issue, and although the probability-weighted consequences of severe accidents are SMALL for all plants, severe accidents remain a Category 2 issue to the extent that only the alternatives to mitigate severe accidents must be considered for all plants that have not previously considered such alternatives.

In addition, it is reasonable that in license renewal applications, the impacts from reactor accidents at full power, including internal and external events, should continue to be considered in assessing SAMAs. The impacts of all other new information do not contribute sufficiently to the environmental impacts to warrant their inclusion in the SAMA analysis since the likelihood of finding cost-effective plant improvements is small. Alternatives to mitigate severe accidents still must be considered for all plants that have not considered such alternatives. Table E-19 provides a summary of the conclusions discussed above.

Table E-19. Summary of Conclusions

Topic (Section)	Conclusions
New Internal Events Information (Section E.3.1)	New information on the risk and environmental impacts of severe accidents caused by internal events indicates that PWR and BWR CDFs are generally comparable to or less than those forming the basis of the 1996 GEIS. In some cases, these differences are significant (approaching one order of magnitude). Comparison of population dose from newer assessments illustrates a reduction in impact by a factor of 5 to 100 when compared to older assessments, and an additional factor of 2 to 4 due to the conservatism built into the 1996 GEIS values. This would also mean that contamination of open bodies of water and economic impacts would, in most cases, be significantly less. Additionally, the likelihood of basemat melt-through accidents is less than that used in the analysis supporting the 1996 GEIS.
Consideration of External Events (Section E.3.2)	The 1996 GEIS did not quantitatively consider severe accidents initiated by external events in assessing environmental impacts. When the environmental impacts of external events are considered, they can be comparable to those from internal events; however, they are generally lower than the estimates used in the 1996 GEIS for internal events. This conclusion would also apply to the contamination of open bodies of water and groundwater and economic impacts.

Table E-19. (cont.)

Topic (Section)	Conclusions
New Source Term Information (Section E.3.3)	More recent source term information indicates that the timing from dominant severe accident sequences, as quantified in NUREG/CR-6295, is comparable to the analysis forming the basis of the 1996 GEIS. In most cases, the release frequencies and release fractions are significantly lower for the more recent estimate. Thus, the environmental impacts used as the basis for the 1996 GEIS are higher than the impacts that would be estimated using the more recent source term information.
Power Uprates (Section E.3.4)	Based on a comparison of the change in LERF for extended power uprates, a small to moderate increase in environmental impacts results from the increase in operating power level.
Higher Fuel Burnup (Section E.3.5)	Increased peak fuel burnup from 42 to 75 GWd/MT for PWRs, and 60 to 75 GWd/MT for BWRs, is estimated to result in small to moderate increases in the environmental impacts in the event of a severe accident.
Consideration of Low Power and Shutdown Events (Section E.3.6)	The environmental impacts from accidents at low power and shutdown conditions are generally comparable to those from accidents at full power when comparing the values in NUREG/CR-6143 and NUREG/CR-6144 to those in NUREG-1150. Even so, the 1996 GEIS estimates of the environmental impact of severe accidents bound the potential impacts from accidents at low power and shutdown. Finally, as cited above and discussed in SECY-97-168, industry initiatives taken during the early 1990s have also contributed to the improved safety of low power and shutdown operation.
Consideration of Spent Fuel Pool Accidents (Section E.3.7)	The environmental impacts from accidents at SFPs (as quantified in NUREG-1738) can be comparable to those from reactor accidents at full power (as estimated in NUREG-1150). Subsequent analyses performed, and mitigative measures employed, since 2001 have further lowered the risk of this class of accidents. In addition, the conservative estimates from NUREG-1738 are much less than the impacts from full power reactor accidents that are estimated in the 1996 GEIS.
Use of BEIR VII Risk Coefficient (Section E.3.8)	Use of newer risk coefficients such as in BEIR VII is expected to have a small impact on the results presented in the 1996 GEIS.
Uncertainties (Section E.3.9)	The impact and magnitude of uncertainties, as estimated in the 1996 GEIS, bound the uncertainties introduced by the new information and considerations.
SAMAs (Section E.4)	The current process and scope of SAMA analysis are sufficient for determining the need for additional mitigative measures.
Summary and Conclusion (Section E.5)	Given the new and updated information, the reduction in estimated environmental impacts from the use of new internal event and source term information outweighs any increases from the consideration of external events, power uprates, higher fuel burnup, low power and shutdown risk, and SFP risk.

E.6 References

10 CFR Part 50. *Code of Federal Regulations*, Title 10, *Energy*, Part 50, "Domestic Licensing of Production and Utilization Facilities."

10 CFR Part 51. *Code of Federal Regulations*, Title 10, *Energy,* Part 51, "Environmental Protection Regulations for Domestic Licensing and Related Regulatory Functions."

10 CFR Part 54. *Code of Federal Regulations*, Title 10, *Energy*, Part 54, "Requirements for Renewal of Operating Licenses for Nuclear Power Plants."

10 CFR Part 72. *Code of Federal Regulations*, Title 10, *Energy*, Part 72, "Licensing Requirements for the Independent Storage of Spent Nuclear Fuel, High-Level Radioactive Waste, and Reactor-Related Greater Than Class C Waste."

10 CFR Part 73. *Code of Federal Regulations*, Title 10, *Energy*, Part 73, "Physical Protection of Plants and Materials."

10 CFR Part 100. *Code of Federal Regulations*, Title 10, *Energy*, Part 100, "Reactor Site Criteria."

61 FR 28467. U.S. Nuclear Regulatory Commission. Environmental Review for Renewal of Nuclear Power Plant Operating Licenses. June 5, 1996.

International Commission on Radiological Protection (ICRP). 1991. *1990 Recommendations of the International Commission on Radiological Protection*. ICRP Publication 60, Pergamon Press, Oxford, United Kingdom.

National Environmental Policy Act (NEPA). 1969.

National Research Council. 1972. *The Effects on Populations of Exposure to Low Levels of Ionizing Radiation (BEIR I)*. National Academies Press, Washington, D.C.

National Research Council. 1980. *The Effects on Populations of Exposure to Low Levels of Ionizing Radiation (BEIR III)*. National Academies Press, Washington, D.C.

National Research Council. 1990. *Health Effects of Exposure to Low Levels of Ionizing Radiation (BEIR V)*. National Academies Press, Washington, D.C.

National Research Council. 2006a. *Health Risks from Exposure to Low Levels of Ionizing Radiation, BEIR VII.* Committee on the Biological Effects of Ionizing Radiation (BEIR). National Academies Press, Washington, D.C.

National Research Council. 2006b. *Safety and Security of Commercial Spent Nuclear Fuel Storage*. National Academies Press, Washington, D.C.

Nuclear Energy Institute (NEI). 2005. *SAMA Analysis Guidance Document.* Revision A, NEI-05-01. November.

Nuclear Management and Resources Council (NUMARC). 1991. *Guidelines for Industry Actions to Assess Shutdown Management.* NUMARC 91-06. December.

U.S. Nuclear Regulatory Commission (NRC). 1975. *Reactor Safety Study—An Assessment of Accident Risks in U.S. Commercial Nuclear Power Plants.* WASH-1400, NUREG-75/014. October.

U.S. Nuclear Regulatory Commission (NRC). 1978. *Liquid Pathway Generic Study.* NUREG-0440. February.

U.S. Nuclear Regulatory Commission (NRC). 1982. *The Development of Severe Reactor Accident Source Terms: 1957–1981.* NUREG-0773. November.

U.S. Nuclear Regulatory Commission (NRC). 1988. *Assessment of the Use of Extended Burn-up Fuel in Light Water Power Reactors.* NUREG/CR-5009. February.

U.S. Nuclear Regulatory Commission (NRC). 1989. *Regulatory Analysis for the Resolution of Generic Issue 82.* NUREG-1353. April.

U.S. Nuclear Regulatory Commission (NRC). 1990a. *Evaluation of Severe Accident Risks: Surry Unit 1.* NUREG/CR-4551 Volume 3, Revision 1. October.

U.S. Nuclear Regulatory Commission (NRC). 1990b. *Severe Accident Risk: An Assessment for Five U.S. Nuclear Power Plants.* NUREG-1150. December.

U.S. Nuclear Regulatory Commission (NRC). 1992. *Integrated Risk Assessment for the LaSalle Unit 2 Nuclear Power Plant.* NUREG/CR-5305. August.

U.S. Nuclear Regulatory Commission (NRC). 1993. *Shutdown and Low Power Operation at Commercial NPPs in the United States.* NUREG-1449. September.

U.S. Nuclear Regulatory Commission (NRC). 1995a. *Evaluation of Potential Severe Accidents During Low Power and Shutdown Operations at Surry, Unit 1.* NUREG/CR-6144. May.

U.S. Nuclear Regulatory Commission (NRC). 1995b. *Evaluation of Potential Severe Accidents During Low Power Shutdown Operation at Grand Gulf, Unit 1.* NUREG/CR-6143. July.

U.S. Nuclear Regulatory Commission (NRC). 1996. *Generic Environmental Impact Statement for License Renewal of Nuclear Plants.* NUREG-1437, Vols. 1 and 2. Office of Nuclear Reactor Regulation, Washington, D.C.

U.S. Nuclear Regulatory Commission (NRC). 1997a. *Regulatory Analysis Technical Evaluation Handbook.* NUREG/BR-0184. January.

U.S. Nuclear Regulatory Commission (NRC). 1997b. *Reassessment of Selected Factors Affecting Siting of Nuclear Power Plants.* NUREG/CR-6295. February.

U.S. Nuclear Regulatory Commission (NRC). 1997c. *Issuance for Public Comment of Proposed Rulemaking Package for Shutdown and Fuel Storage Pool Operation.* SECY-97-168. July 30.

U.S. Nuclear Regulatory Commission (NRC). 1997d. *Issuance for Public Comment of Proposed Rulemaking Package for Shutdown and Fuel Storage Pool Operation.* SRM-SECY-97-168. December 11.

U.S. Nuclear Regulatory Commission (NRC). 1999. *Generic Environmental Impact Statement for License Renewal of Nuclear Plants, Main Report*, "Section 6.3—Transportation, Table 9.1, Summary of Findings on NEPA Issues for License Renewal of Nuclear Power Plants, Final Report." NUREG-1437, Vol. 1, Addendum 1. Office of Nuclear Reactor Regulation, Washington, D.C.

U.S. Nuclear Regulatory Commission (NRC). 2001a. *Environmental Impact of Extending Fuel Burn-up Above 60 GWd/MT.* NUREG/CR-6703. January.

U.S. Nuclear Regulatory Commission (NRC). 2001b. *Technical Study of Spent Fuel Pool Accident Risk and Decommissioning Nuclear Power Plants.* NUREG-1738. February.

U.S. Nuclear Regulatory Commission (NRC). 2001c. *Generic Aging Lessons Learned Report.* NUREG-1801. July.

U.S. Nuclear Regulatory Commission (NRC). 2003. *Perspectives Gained from the Individual Plant Examination of External Events (IPEEE).* NUREG-1742. April.

U.S. Nuclear Regulatory Commission (NRC). 2004. *Protecting Our Nation—Since 9-11-01.* NUREG/BR-0314. September.

U.S. Nuclear Regulatory Commission (NRC). 2005a. Letter from Nils J. Diaz (NRC Chairman) to Senator Pete V. Domenici (Chairman, Subcommittee on Energy and Water Development), dated March 14, 2005.

U.S. Nuclear Regulatory Commission (NRC). 2005b. *Staff Review of the National Academies Study of the Health Risks from Exposure to Low Levels of Ionizing Radiation (BEIR VII).* SECY-05-0202. October 29.

U.S. Nuclear Regulatory Commission (NRC). 2006a. *Frequently Asked Questions on License Renewal of Nuclear Power Reactors.* NUREG-1850. March.

U.S. Nuclear Regulatory Commission (NRC). 2006b. *Mitigation of Spent Fuel Loss-of-Coolant Inventory Accidents and Extension of Reference Plant Analyses to Other Spent Fuel Pools.* Sandia Letter Report, Revision. November 2.

U.S. Nuclear Regulatory Commission (NRC). 2007a. Memorandum from Stewart Bailey (NRC) to James Scarola (Brunswick Steam Electric Plant) entitled, "Brunswick Steam Electric Plant, Units 1 and 2—Conforming License Amendments to Incorporate the Mitigation Strategies Required by Section B.5.b. of Commission Order EA-02-026 and the Radiological Protection Mitigation Strategies Required by Commission Order EA-06-137 (TAC Nos. MD4516 and MD4517)." August 9.

U.S. Nuclear Regulatory Commission (NRC). 2007b. *2007–2008 Information Digest.* NUREG-1350, Vol. 19. August.

U.S. Nuclear Regulatory Commission (NRC). 2007c. *A Prioritization of Generic Safety Issues.* NUREG-0933, Supplement 31. September.

U.S. Nuclear Regulatory Commission (NRC). 2007d. Denial of Petition for Rulemaking PRM-51-11. *Federal Register* 72(240):71083. December 14.

U.S. Nuclear Regulatory Commission (NRC). 2008a. "State-of-the-Art Reactor Consequence Analyses (SOARCA) Project Accident Analysis," Randall Gauntt and Charles Tinkler, presented at the 2007 Regulatory Information Conference (RIC), Rockville, Maryland. March 11.

U.S. Nuclear Regulatory Commission (NRC). 2008b. Letter from M. Gamberoni (NRC) to J. Pollock (Entergy Nuclear Operations) entitled, "Indian Point Nuclear Generating Units 1 & 2—NRC Inspection Report Nos. 05000003/2007010 and 05000247/2007010." May 13.

U.S. Nuclear Regulatory Commission (NRC). 2008c. *Industry Performance Data*. Available URL: http://nrcoe.inel.gov/results/index.cfm?fuseaction=IndustryPerf.showMenu (Accessed August 4, 2008).

U.S. Nuclear Regulatory Commission (NRC). 2008d. Denial of Petition for Rulemaking PRM-51-10 and PRM-51-12. *Federal Register* 73(154):46204. August 8.

U.S. Nuclear Regulatory Commission (NRC). 2008e. *Resolution of Generic Safety Issues.* NUREG-0933, Main Report and Supplements 1–32. August.

U.S. Nuclear Regulatory Commission (NRC). 2011. *Recommendations for Enhancing Reactor Safety in the 21st Century, the Near-Term Task Force Review of Insights from the Fukushima Dai-ichi Accident.*" July 12. ADAMS Accession No. ML111861807.

U.S. Nuclear Regulatory Commission (NRC). 2013. *Preparation of Environmental Reports for Nuclear Power Plant License Renewal Applications.* Regulatory Guide 4.2, Supplement 1, Revision 1.

Appendix F

Laws, Regulations, and Other Requirements

Appendix F

Laws, Regulations, and Other Requirements

F.1 Introduction

This appendix presents a brief discussion of Federal and State laws, regulations, and other requirements that may be affected by the renewal and continued operation of U.S. Nuclear Regulatory Commission (NRC)-licensed nuclear power plants. It provides additional information about environmental laws and regulations, applicable to license renewal, that were introduced in Chapter 3, "Affected Environment." These include Federal and State laws, regulations, and other requirements designed to protect the environment, including land and water use, air quality, aquatic resources, terrestrial resources, radiological impacts, waste management, chemical impacts, and socioeconomic conditions.

Applicable Federal and State laws and regulations presented in this part include:

(1) laws and regulations that could require the NRC or the applicant to undergo a *new* authorization or consultation process with Federal or State agencies outside the NRC or,

(2) laws and executive orders that could require the NRC or the applicant to *renew* authorizations currently granted or hold additional consultations with Federal or State agencies outside the NRC.

This appendix is provided as a basic overview to assist the applicant in identifying environmental and natural resources laws that may affect the license renewal process. The descriptions of each of the laws, regulations, executive orders, and other directives are general in nature and are not intended to provide a comprehensive analysis or explanation of any of the items listed. In addition, the list itself is not intended to be comprehensive, and an applicant for license renewal is reminded that a variety of additional Federal, State, or local requirements may apply to a license renewal application for a particular plant site.

Section F.3 identifies Federal laws and regulations applicable to license renewal. Section F.4 discusses Executive Orders. Section F.5 identifies applicable NRC regulations. Section F.6 discusses State laws, regulations, and agreements. Section F.7 discusses emergency management and response laws, regulations, and executive orders. Section F.8 discusses consultations with agencies and Federally recognized American Indian Nations. These regulatory requirements address issues such as protection of public health and the environment, worker safety, historic and cultural resources, and emergency planning.

F.2 Background

The NRC is required to ensure that licensed nuclear power plants are operated in a manner that ensures the protection of public health and safety and the environment.

There are a number of Federal laws and regulations that affect environmental protection, health, safety, compliance, and/or consultation at every NRC-licensed nuclear power plant. In addition, certain Federal environmental requirements have been delegated to State authorities for enforcement and implementation. Furthermore, States have also enacted laws to protect public health and safety and the environment. It is NRC's policy to make sure nuclear power plants are operated in a manner that ensures the protection of public health and safety and the environment through compliance with applicable Federal and State laws, regulations, and other requirements.

F.3 Federal Laws and Regulations

The requirements that may be applicable to the operation of NRC-licensed nuclear power plants encompass a broad range of Federal laws and regulations, addressing environmental, historic and cultural, health and safety, transportation, and other concerns. Generally, these laws and regulations are relevant to how the work involved in performing a proposed action would be conducted to protect workers, the public, and environmental resources. Some of these laws and regulations require permits or consultation with other Federal agencies or state, tribal, or local governments. The Federal laws and regulations that are identified and briefly discussed in this section are presented in alphabetical order.

American Indian Religious Freedom Act (AIRFA) of 1978 (42 United States Code [USC] 1996)—The American Indian Religious Freedom Act protects Native Americans' rights of freedom to believe, express, and exercise traditional religions.

Antiquities Act of 1906, as amended (16 USC 431–433)—The Antiquities Act protects historic and prehistoric ruins, monuments, and antiquities, including paleontological resources, on Federally controlled lands from appropriation, excavation, injury, and destruction without permission.

Archaeological and Historic Preservation Act of 1960, as amended (16 USC 469 et seq.)—The Archaeological and Historic Preservation Act establishes procedures for preserving historical and archeological resources. Analysis of environmental compliance included assessing the energy alternatives for possible impacts on prehistoric, historic, and traditional cultural resources.

Archaeological Resources Protection Act of 1979, as amended (16 USC 470aa et seq.)—The Archaeological Resources Protection Act requires a permit for any excavation or removal of archaeological resources from Federal or American Indian lands. Excavations must be undertaken for the purpose of furthering archaeological knowledge in the public interest and resources removed are to remain the property of the United States. Consent must be obtained from the American Indian Tribe or the Federal agency having authority over the land, on which a resource is located, before issuance of a permit. The permit must contain terms and conditions requested by the Tribe or Federal agency.

Atomic Energy Act of 1954 (42 USC 2011 et seq.)—The 1954 Atomic Energy Act (AEA), as amended, and the Energy Reorganization Act of 1974 (42 USC 5801 et seq.) gives the NRC the licensing and regulatory authority for nuclear energy uses within the commercial sector. It gives NRC responsibility for licensing and regulating commercial uses of atomic energy and allows the NRC to establish dose and concentration limits for protection of workers and the public for activities under NRC jurisdiction. NRC implements its responsibilities under the AEA through regulations set forth in Title 10 of the CFR.

Bald and Golden Eagle Protection Act of 1940, as amended (16 USC 668–668d)—The Bald and Golden Eagle Protection Act makes it unlawful to take, pursue, molest, or disturb bald and golden eagles, their nests, or their eggs anywhere in the United States. The U.S. Fish and Wildlife Service (USFWS) may issue take permits to individuals, government agencies, or other organizations to authorize limited, non-purposeful disturbance of eagles, in the course of conducting lawful activities such as operating utilities or conducting scientific research.

Clean Air Act of 1970, as amended (42 USC 7401 et seq.)—The Clean Air Act (CAA) is intended to "protect and enhance the quality of the nation's air resources so as to promote the public health and welfare and the productive capacity of its population." The CAA establishes regulations to ensure maintenance of air quality standards and authorizes individual States to manage permits. Section 118 of the CAA requires each Federal agency, with jurisdiction over properties or facilities engaged in any activity that might result in the discharge of air pollutants, to comply with all Federal, State, inter-State, and local requirements with regard to the control and abatement of air pollution. Section 109 of the CAA directs the U.S. Environmental Protection Agency (EPA) to set National Ambient Air Quality Standards (NAAQS) for criteria pollutants. The EPA has identified and set NAAQS for the following criteria pollutants: particulate matter, sulfur dioxide, carbon monoxide, ozone, nitrogen dioxide, and lead.

Section 111 of the CAA requires establishment of national performance standards for new or modified stationary sources of atmospheric pollutants. Section 160 of the CAA requires that specific emission increases must be evaluated prior to permit approval in order to prevent significant deterioration of air quality. Section 112 requires specific standards for release of hazardous air pollutants (including radionuclides). These standards are implemented through plans developed by each State and approved by the EPA. The CAA requires sources to meet standards and obtain permits to satisfy those standards. Nuclear power plants may be required to comply with the CAA Title V, Sections 501–507, for sources subject to new source performance standards or sources subject to National Emission Standards for Hazardous Air Pollutants. Emissions of air pollutants are regulated by the EPA in 40 CFR Parts 50 to 99.

Clean Water Act (33 USC 1251 et seq.)—The Clean Water Act (CWA; formerly the Federal Water Pollution Control Act) was enacted to "restore and maintain the chemical, physical, and biological integrity of the Nation's water." The Act requires all branches of the Federal Government, with jurisdiction over properties or facilities engaged in any activity that might result in a discharge or runoff of pollutants to surface waters, to comply with Federal, State, inter-State, and local requirements.

As authorized by the Clean Water Act, the National Pollutant Discharge Elimination System (NPDES) permit program controls water pollution by regulating point sources that discharge pollutants into waters of the United States. The NPDES program requires all facilities that discharge pollutants from any point source into waters of the United States obtain an NPDES permit. An NPDES permit is developed with two levels of controls: technology-based limits and water quality-based limits. NPDES permit terms may not exceed 5 years, and the applicant must reapply at least 180 days prior to the permit expiration date. A nuclear power plant may also participate in the NPDES General Permit for Industrial Stormwater due to stormwater runoff from industrial or commercial facilities to waters of the United States. EPA is authorized under the CWA to directly implement the NPDES program; however, EPA has authorized many States to implement all or parts of the national program. Section 401 of the CWA requires States to certify that the permitted discharge would comply with all limitations necessary to meet established State water quality standards, treatment standards, or schedule of compliance.

The U.S. Army Corps of Engineers (USACE) is the lead agency for enforcement of CWA wetland requirements (33 CFR Part 320). Under Section 401 of the CWA, the EPA or a delegated State agency has the authority to review and approve, condition, or deny all permits or licenses that might result in a discharge to waters of the State, including wetlands.

A Section 404 permit would need to be obtained from the USACE before implementing any action, such as earthmoving activities and certain erosion controls, which could disturb wetlands. Federal and State permits/certification are obtained using the same form and permit applications for activities affecting waterways and wetlands are reviewed by the USACE in consultation with the USFWS, the Soil Conservation Service, the EPA, and the delegated State agency.

Coastal Zone Management Act of 1972, as amended (16 USC 1451 et seq.)—Congress enacted the *Coastal Zone Management Act* (CZMA) in 1972 to address the increasing pressures of over-development upon the nation's coastal resources. The National Oceanic and Atmospheric Administration (NOAA) administers the Act. The CZMA encourages States to preserve, protect, develop, and, where possible, restore or enhance valuable natural coastal resources such as wetlands, floodplains, estuaries, beaches, dunes, barrier islands, and coral reefs, as well as the fish and wildlife using those habitats. Participation by States is voluntary. To encourage States to participate, the CZMA makes Federal financial assistance available to any coastal State or territory, including those on the Great Lakes, which are willing to develop and implement a comprehensive coastal management program.

Comprehensive Environmental Response, Compensation, and Liability Act as amended by the Superfund Amendments and Reauthorization Act (42 USC 9601 et seq.)—The Comprehensive Environmental Response, Compensation, and Liability Act (CERCLA) includes an emergency response program to respond to a release of a hazardous substance to the environment. Releases of source, byproduct, or special nuclear material from a nuclear incident are excluded from CERCLA requirements if the releases are subject to the financial protection requirements of the AEA. CERCLA is intended to provide a response to, and cleanup of, environmental problems that are not covered adequately by the permit programs of the many other environmental laws, including the CAA, CWA, Safe Drinking Water Act, Marine Protection, Research, and Sanctuaries Act; *Resource Conservation and Recovery Act*, and AEA. Under Section 120 of CERCLA, each department, agency, and instrumentality (e.g., a municipality) of the United States is subject to, and must comply with, CERCLA in the same manner as any nongovernmental entity (except for requirements for bonding, insurance, financial responsibility, or applicable time period). Under CERCLA, the EPA would have the authority to regulate hazardous substances at a facility in the event of a release or a "substantial threat of a release" of those materials. Releases greater than reportable quantities would be reported to the National Response Center. Assessment of alternatives for environmental compliance includes consideration of whether hazardous substances, in reportable quantity amounts, could be present at power plants during the license renewal term.

Emergency Planning and Community Right-to-Know Act of 1986 (42 USC 11001 et seq.) (also known as "SARA Title III")—The Emergency Planning and Community Right-to-Know Act of 1986 (EPCRA), which is the major amendment to CERCLA (42 USC 9601*)*, establishes the requirements for Federal, State, and local governments, Indian Tribes, and industry regarding emergency planning and "Community Right-to-Know" reporting on hazardous and toxic chemicals. The "Community Right-to-Know" provisions increase the public's knowledge and access to information on chemicals at individual facilities, their uses, and releases into the environment. States and communities working with facilities can use the information to improve chemical safety and protect public health and the environment. This Act requires emergency planning and notice to communities and government agencies concerning the presence and release of specific chemicals. The EPA implements this Act under regulations found in 40 CFR Parts 355, 370, and 372.

Endangered Species Act of 1973 (16 USC 1531–1544)—The Endangered Species Act (ESA) was enacted to prevent the further decline of endangered and threatened species and to restore those species and their critical habitats. Section 7 of the Act requires Federal agencies to consult with the USFWS or the National Marine Fisheries Service (NMFS) for Federal actions that may affect listed species or designated critical habitats.

Environmental Standards for Uranium Fuel Cycle (40 CFR Part 190, Subpart B)—These regulations establish maximum doses to the body or organs of members of the public as a result of normal operational releases from uranium fuel cycle activities, including uranium enrichment. These regulations were promulgated by the EPA under the authority of the AEA, as amended, and have been incorporated by reference in the NRC regulations in 10 CFR 20.1301(e).

Federal Insecticide, Fungicide, and Rodenticide Act, as amended (7 USC 135 et seq.)—The *Federal Insecticide, Fungicide, and Rodenticide Act,* as amended, by the Federal Environmental Pesticide Control Act and subsequent amendments, requires the registration of all new pesticides with the EPA before they are used in the United States. Manufacturers are required to develop toxicity data for their pesticide products. Toxicity data may be used to determine permissible discharge concentrations for an NPDES permit.

Fish and Wildlife Conservation Act of 1980 (16 USC 2901 et seq.)—The *Fish and Wildlife Conservation Act* (FWCA) provides Federal technical and financial assistance to States for the development of conservation plans and programs for nongame fish and wildlife. FWCA conservation plans identify significant problems that may adversely affect non-game fish and wildlife species and their habitats and appropriate conservation actions to protect the identified species. The Act also encourages Federal agencies to conserve and promote the conservation of nongame fish and wildlife and their habitats.

Fish and Wildlife Coordination Act of 1934, as amended (16 USC 661–666e)—The Fish and Wildlife Coordination Act requires Federal agencies that construct, license, or permit water resource development projects to consult with the USFWS (or NMFS, when applicable) and State wildlife resource agencies for any project that involves an impoundment of more than 10 acres (4 hectares), diversion, channel deepening, or other water body modification regarding the impacts of that action to fish and wildlife and any mitigative measures to reduce adverse impacts.

Hazardous Materials Transportation Act of 1975, as amended (49 USC 1801 et seq.)—The Hazardous Materials Transportation Act regulates the transportation of hazardous material (including radioactive material) in and between States. According to the Act, States may regulate the transport of hazardous material as long as their regulation is consistent with the Act or the U.S. Department of Transportation (USDOT) regulations provided in 49 CFR Parts 171 through 177. Other regulations regarding packaging for transportation of radionuclides are contained in 49 CFR Part 173, Subpart I.

Low-Level Radioactive Waste Policy Act of 1980, as amended (42 USC 2021 et seq.)—The Low-Level Radioactive Waste Policy Act amended the AEA to improve the procedures for the implementation of compacts providing for the establishment and operation of regional low-level radioactive waste disposal facilities. It also allows for Congress to grant consent for certain inter-State compacts. The amended Act sets forth the responsibilities for disposal of low-level waste by States or inter-State compacts. The Act states the amount of waste that certain low-level waste recipients can receive over a set time period. The amount of low-level radioactive waste generated from both pressurized and boiling water reactor types is allocated over a transition period until a local waste facility is operational.

Magnuson-Stevens Fishery Conservation and Management Act, as amended (16 USC 1801-1884)—The Magnuson-Stevens Fishery Conservation and Management Act (MSA) governs marine fisheries management in U.S. Federal waters. The Act created eight regional fishery management councils and includes measures to rebuild overfished fisheries, protect essential fish habitat, and reduce bycatch. Under Section 305 of the Act, Federal agencies are required to consult with NMFS for any Federal actions that may adversely affect essential fish habitat.

Marine Mammal Protection Act of 1972 (16 USC 1361 et seq.)—The *Marine Mammal Protection Act* (MMPA) was enacted to protect and manage marine mammals and their products (e.g., the use of hides and meat). The primary authority for implementing the Act belongs to the USFWS and NMFS. The USFWS manages walruses, polar bears, sea otters, dugongs, marine otters, and the West Indian, Amazonian, and West African manatees. The NMFS manages whales, porpoises, seals, and sea lions. The two agencies may issue permits under MMPA Section 104 (16 USC 1374) to persons, including Federal agencies, that authorize the taking or importing of specific species of marine mammals.

After the Secretary of the Interior or the Secretary of Commerce approves a State's program, the State can take over responsibility for managing one or more marine mammals. The MMPA also established a Marine Mammal Commission whose duties include reviewing laws and international conventions relating to marine mammals, studying the condition of these mammals, and recommending steps to Federal officials (e.g., listing a species as endangered) that should be taken to protect marine mammals. Federal agencies are directed by MMPA

Section 205 (16 USC 1405) to cooperate with the commission by permitting it to use their facilities or services.

Migratory Bird Treaty Act of 1918, as amended (16 USC 703 et seq.)—The Migratory Bird Treaty Act is intended to protect birds that have common migration patterns between the United States and Canada, Mexico, Japan, and Russia. The Act stipulates that, except as permitted by regulations, it is unlawful at any time, by any means, or in any manner to pursue, hunt, take, capture, or kill any migratory bird.

National Environmental Policy Act of 1969, as amended (42 USC 4321 et seq.)—The National Environmental Policy Act (NEPA) requires Federal agencies to integrate environmental values into their decision-making process by considering the environmental impacts of proposed Federal actions and reasonable alternatives to those actions. NEPA establishes policy, sets goals (in Section 101), and provides means (in Section 102) for carrying out the policy. Section 102(2) contains action-forcing provisions to ensure that Federal agencies follow the letter and spirit of the Act. For major Federal actions significantly affecting the quality of the human environment, Section 102(2)(C) of NEPA requires Federal agencies to prepare a detailed statement that includes the environmental impacts of the proposed action and other specified information. This generic environmental impact statement (GEIS) has been prepared in accordance with NEPA requirements and NRC regulations (10 CFR Part 51) for implementing NEPA to ensure compliance with Section 102(2).

National Historic Preservation Act of 1966, as amended (16 USC 470aa et seq.)—The National Historic Preservation Act (NHPA) was enacted to create a national historic preservation program, including the *National Register of Historic Places* and the Advisory Council on Historic Preservation. Section 106 of the Act requires Federal agencies to take into account the effects of their undertakings on historic properties. The Advisory Council on Historic Preservation regulations implementing Section 106 of the Act, are found in 36 CFR Part 800. The regulations call for public involvement in the Section 106 consultation process, including Indian Tribes and other interested members of the public, as applicable.

Native American Graves Protection and Repatriation Act of 1990 (25 USC 3001)—The Native American Graves Protection and Repatriation Act establishes provisions for the treatment of inadvertent discoveries of American Indian remains and cultural objects. When discoveries are made during ground-disturbing activities, the activity in the area must immediately stop, and reasonable protective efforts, proper notifications, and appropriate disposition of the discovered items must be pursued.

Noise Control Act of 1972 (42 USC 4901 et seq.)—The Noise Control Act delegates the responsibility of noise control to State and local governments. Commercial facilities are required to comply with Federal, State, inter-State, and local requirements regarding noise control. Section 4 of the Noise Control Act directs Federal agencies to carry out programs in their jurisdictions "to the fullest extent within their authority" and in a manner that furthers a national policy of promoting an environment free from noise that jeopardizes health and welfare.

Nuclear Regulatory Commission License Termination Rule (10 CFR Part 20, Subpart E)— The AEA assigns NRC the responsibility for licensing and regulating commercial uses of atomic energy. When a licensed facility has completed its mission, the facility must meet standards for cleanup in order to terminate its license. The License Termination Rule establishes that NRC will consider a site acceptable for unrestricted use if the residual radioactivity, that is distinguishable from background radiation, results in a total effective dose equivalent (TEDE) to an average member of the critical group does not exceed 25 millirem per year, including that from groundwater sources of drinking water, and the residual radioactivity has been reduced to levels that are as low as reasonably achievable (ALARA). The critical group is the group of individuals reasonably expected to receive the greatest exposure to residual radioactivity for any applicable set of circumstances.

The License Termination Rule also provides for land use restrictions or other types of institutional controls to allow terminating NRC licenses and releasing sites under restricted conditions if decommissioning criteria for unrestricted use cannot be met. Plus, the License Termination Rule establishes alternate criteria for license termination if the licensee provides assurance that public health and safety would continue to be protected, and that it is unlikely that the dose from all manmade sources combined, other than medical, would be more than 100 millirem per year.

Nuclear Waste Policy Act of 1982 (42 USC 10101 et seq.)—The *Nuclear Waste Policy Act* provides for the research and development of repositories for the disposal of high-level radioactive waste, spent nuclear fuel, and low-level radioactive waste. Title I includes the provisions for the disposal and storage of high-level radioactive waste and spent nuclear fuel. Subtitle A of Title I delineates the requirements for site characterization and construction of the repository and the participation of States and other local governments in the selection process. Subtitles B, C, and D of Title I deal with the specific issues for interim storage, monitored retrievable storage, and low-level radioactive waste.

Occupational Safety and Health Act of 1970 (29 USC 651 et seq.)—The Occupational Safety and Health Act establishes standards to enhance safe and healthy working conditions in places of employment throughout the United States. The Act is administered and enforced by the Occupational Safety and Health Administration (OSHA), a U.S. Department of Labor agency. Employers who fail to comply with OSHA standards can be penalized by the Federal government. The Act allows States to develop and enforce OSHA standards if such programs have been approved by the Secretary of Labor.

Pollution Prevention Act of 1990 (42 USC 13101 et seq.)—The Pollution Prevention Act establishes a national policy for waste management and pollution control that focuses first on source reduction, then on environmental issues, safe recycling, treatment, and disposal.

Resource Conservation and Recovery Act as amended by the Hazardous and Solid Waste Amendments (42 USC 6901 et seq.)—The Resource Conservation and Recovery Act (RCRA) requires the EPA to define and identify hazardous waste; establish standards for its transportation, treatment, storage, and disposal; and require permits for persons engaged in hazardous waste activities. Section 3006 (42 USC 6926) allows States to establish and administer these permit programs with EPA approval. EPA regulations implementing the RCRA are found in 40 CFR Parts 260 through 283. Regulations imposed on a generator or on a treatment, storage, and/or disposal facility vary according to the type and quantity of material or waste generated, treated, stored, and/or disposed. The method of treatment, storage, and/or disposal also impacts the extent and complexity of the requirements.

Safe Drinking Water Act of 1974 (42 USC 300(f) et seq.)—The Safe Drinking Water Act (SDWA) was enacted to protect the quality of public water supplies and sources of drinking water and establishes minimum national standards for public water supply systems in the form of maximum contaminant levels (MCLs) for pollutants, including radionuclides. Other programs established by the SDWA include the Sole Source Aquifer Program, the Wellhead Protection Program, and the Underground Injection Control Program. In addition, the Act provides underground sources of drinking water with protection from contaminated releases and spills.

If a nuclear power plant is located within an area designated as a sole source aquifer pursuant to Section 1424(e) of the SDWA, the supplemental EIS would be subject to EPA review. If the EPA review raises concerns that plant operations are not protective of groundwater quality, specific mitigation recommendations or additional pollution prevention requirements may be required.

Toxic Substances Control Act (15 USC 2601 et seq.)—The Toxic Substances Control Act (TSCA) regulates the manufacture, processing, distribution, and use of certain chemicals not regulated by RCRA or other statutes, including asbestos-containing material and polychlorinated biphenyls. Any TSCA-regulated waste removed from structures (e.g., polychlorinated biphenyls-contaminated capacitors or asbestos) or discovered during the implementation phase (e.g., contaminated media) would be managed in compliance with TSCA requirements in 40 CFR Part 761.

F.4 Executive Orders

Executive Orders establish policies and requirements for Federal agencies. Executive Orders do not have the force of law or regulation. Generally, Executive Orders are applicable to most Federal agencies, although they may or may not be binding upon independent regulatory agencies such as the NRC.

Executive Order 11514, Protection and Enhancement of Environmental Quality—This Order (regulated by 40 CFR Parts 1500 through 1508) requires Federal agencies to continually monitor and control their activities to: (1) protect and enhance the quality of the environment, and (2) develop procedures to ensure the fullest practicable provision of timely public information and understanding of the Federal plans and programs that may have potential environmental impact so that views of interested parties can be obtained.

Executive Order 11593, Protection and Enhancement of the Cultural Environment—This Order directs Federal agencies to locate, inventory, and nominate qualified properties under their jurisdiction or control to the *National Register of Historic Places*.

Executive Order 11988, Floodplain Management—This Order requires Federal agencies to avoid direct or indirect support of floodplain development whenever there is a practicable alternative. A Federal agency is required to evaluate the potential effects of any actions it may take in a floodplain. Federal agencies are also required to encourage and provide appropriate guidance to applicants to evaluate the effects of their proposals on floodplains prior to submitting applications for Federal licenses, permits, loans, or grants.

Executive Order 11990, Protection of Wetlands—This Order requires Federal agencies to avoid any short or long-term adverse impacts on wetlands, wherever there is a practicable alternative and to provide opportunity for early public review of any plans or proposals for new construction in wetlands. Federal agencies are required to evaluate the potential effects of any actions they may take on wetlands when carrying out their responsibilities (e.g., planning, regulating, and licensing activities). However, this executive order does not apply to the issuance by Federal agencies of permits, licenses, or allocations to private parties for activities involving wetlands on non-Federal property.

Executive Order 12088, Federal Compliance with Pollution Control Standards, as amended by Executive Order 12580, Superfund Implementation—This Order directs Federal agencies to comply with applicable administrative and procedural pollution controls standards established by, but not limited to, the CAA, the Noise Control Act, the CWA, the SDWA, the TSCA, and the RCRA.

Executive Order 12148, Federal Emergency Management—This Order transfers functions and responsibilities associated with Federal emergency management to the Director of the Federal Emergency Management Agency. The Order assigns the Director the responsibility to establish Federal policies for, and to coordinate all civil defense and civil emergency planning, management, mitigation, and assistance functions of, Executive agencies.

Executive Order 12580, Superfund Implementation, as amended by Executive Order 13308—This Order delegates to the heads of Executive Departments and agencies the responsibility of undertaking remedial actions for releases or threatened releases that are not on the National Priorities List, and removal actions, other than emergencies, where the release is from any facility under the jurisdiction or control of Executive Departments and agencies.

Executive Order 12656, Assignment of Emergency Preparedness Responsibilities—This Order assigns emergency preparedness responsibilities to Federal departments and agencies.

Executive Order 12856, Right to Know Laws and Pollution Prevention Requirements—The Order directs Federal agencies to reduce and report toxic chemicals entering any waste stream; improve emergency planning, response, and accident notification; and to meet the requirements of EPCRA.

Executive Order 12898, Federal Actions to Address Environmental Justice in Minority Populations and Low-Income Populations—This Order calls for Federal agencies to address environmental justice in minority populations and low-income populations (59 FR 7629), and directs Federal agencies to identify and address, as appropriate, disproportionately high and adverse health or environmental effects of their programs, policies, and activities on minority populations and low-income populations. In response to this Executive Order, the NRC has issued a final policy statement on the "Treatment of Environmental Justice Matters in NRC Regulatory and Licensing Actions" (69 FR 52040) and environmental justice procedures to be followed in NEPA documents.

Executive Order 12902, Energy Efficiency and Water Conservation at Federal Facilities—This Order requires Federal agencies to develop and implement a program for conservation of energy and water resources. As part of this program, agencies are required to conduct comprehensive facility audits of their energy and water use.

Executive Order 13007, Indian Sacred Sites—This Order directs Federal agencies, to the extent permitted by law and not inconsistent with agency missions, to avoid adverse effects to sacred sites and to provide access to those sites to Native Americans for religious practices. The Order directs agencies to plan projects, to provide protection of, and access to sacred sites to the extent compatible with the project.

Executive Order 13045, Protection of Children from Environmental Health Risks and Safety Risks, as amended by Executive Order 13229, as amended by Executive Order 13296—This Order requires Federal Executive branch agencies to make it a high priority to identify and assess environmental health risks and safety risks that may disproportionately affect children and to ensure that its policies, programs, activities, and standards address disproportionate risks to children that result from environmental health or safety risks.

Executive Order 13101, Greening the Government through Waste Prevention, Recycling, and Federal Acquisition—This Order requires each Federal agency to incorporate waste prevention and recycling in its daily operations and work to increase and expand markets for recovered materials. This Order states that it is national policy to prefer pollution prevention whenever feasible. Pollution that cannot be prevented should be recycled; pollution that cannot be prevented or recycled should be treated in an environmentally safe manner. Disposal should be employed only as a last resort.

Executive Order 13112, Invasive Species—This Order directs Federal agencies to act to prevent the introduction of or to monitor and control, invasive (nonnative) species, to provide for restoration of native species, to conduct research, to promote educational activities, and to exercise care in taking actions that could promote the introduction or spread of invasive species. During the implementation phase, rehabilitation of disturbed areas would be accomplished by reseeding or revegetating areas with native plants and trees.

Executive Order 13123, Greening the Government through Efficient Energy Management—This Order sets goals for agencies to reduce greenhouse gas emissions from facility energy use, reduce energy consumption per gross square foot of facilities, reduce energy consumption per gross square foot or unit of production, expand use of renewable energy, reduce the use of petroleum within facilities, reduce source energy use, and reduce water consumption and associated energy use.

Executive Order 13148, Greening the Government through Leadership in Environmental Management—This Order requires agencies to develop strategies and goals for environmental compliance, right-to-know, and pollution prevention. It requires all Federal facilities to have an environmental management system, requires compliance or environmental management system audits, and requires that Federal Executive Branch agencies comply with the requirements for toxic chemical release reporting in Section 313 of EPCRA.

Executive Order 13175, Consultation and Coordination with Indian Tribal Governments—This Order directs Federal agencies to establish regular and meaningful consultation and collaboration with Tribal governments in the development of Federal policies that have Tribal implications, to strengthen U.S. government-to-government relationships with American Indian Tribes, and to reduce the imposition of unfunded mandates on Tribal governments.

F.5 U.S. Nuclear Regulatory Commission Regulations

The AEA, as amended, allows the NRC to issue licenses for commercial power reactors to operate up to 40 years. This license is based on adherence of the licensee to NRC's regulations which are set forth in Chapter 1 of Title 10 of the CFR. The NRC regulations allow for the renewal of the licenses for up to an additional 20 years beyond the initial licensing period. The renewal of the license depends on the outcome of the NRC's safety and environmental reviews of the commercial power reactor license renewal applications. There are no specific limitations in the AEA or NRC regulations restricting the number of times a license may be renewed. The license renewal process includes a set of requirements, which are designed to assure safe operation and protection of the environment.

The license renewal process includes two reviews: an environmental review and a safety review. The reviews are based on the regulations published in 10 CFR Part 51, for the environmental review and 10 CFR Part 54 for the safety review. These regulations prescribe the format and content of license renewal applications, as well as, the methods and criteria used by NRC staff in evaluating these applications.

The license renewal environmental review relies upon the following regulations and guidance:

- *Code of Federal Regulations*—The scope of the environmental review is based on the regulations provided in 10 CFR Part 51, "Environmental Protection Regulations for Domestic Licensing and Related Regulatory Functions."

- *Preparation of Environmental Reports for License Renewal Applications (Supplement 1 to Regulatory Guide 4.2, Revision 1)*—This document outlines the format and content to be used by the applicant to discuss the environmental aspects of its license renewal application. It also defines the information and analyses the applicant must include in its environmental report submitted as part of the application.

- *Standard Review Plan for Environmental Reviews for Nuclear Power Plants—Supplement 1: Operating License Renewal (NUREG-1555, Supplement 1, Revision 1)*—This document describes how the NRC staff conducts its review of the environmental issues associated license renewal.

- *Generic Environmental Impact Statement for License Renewal of Nuclear Plants* (GEIS) (NUREG-1437, Revision 1)—This document discusses the environmental impacts from license renewal that are common to all or most nuclear power facilities. The GEIS allows the applicant and NRC to focus on environmental issues specific to each site seeking a renewed operating license. The staff's review results in a site-specific supplement to the GEIS for each plant site.

F.6 State Laws, Regulations, and Other Requirements

The AEA authorizes States to establish programs to assume NRC regulatory authority for certain activities (the NRC's Agreement State program). The New York State Department of Labor (NYSDOL) and Department of Environmental Conservation (NYSDEC), for example, have established requirements under this Agreement State Program. NYSDOL has jurisdiction in New York over commercial and industrial uses of radioactive material. Under the New York Agreement State Program, NYSDEC has jurisdiction over discharges of radioactive material to the environment, including releases to the air and water, and the disposal of radioactive wastes in the ground. In addition, States have enacted their own laws to protect public health and safety, and the environment. State laws may supplement or implement various Federal laws for protection of air, water quality, and groundwater. State laws may also address solid waste management programs, locally rare or endangered species, and historic and cultural resources.

In addition, the CWA allows for primary enforcement and administration through State agencies, provided the State program (1) is at least as stringent as the Federal program and (2) conforms to the CWA. The primary CWA mechanism to control water pollution is the requirement that direct dischargers obtain an NPDES permit or, in the case of States where the authority has been delegated from the EPA, a State permit.

One important difference between Federal regulations and certain State regulations is the definition of waters regulated by the State. Certain State regulations may include underground waters, while the CWA only regulates the navigable waters of the United States. For example, a State permit is required under New York State law for all discharges to both surface waters and groundwater.

F.6.1 State Environmental Requirements

Certain environmental requirements, including some discussed earlier, may have been delegated to State authorities for implementation, enforcement, or oversight. Table F.6-1 provides a list of representative State environmental requirements that may affect license renewal applications for nuclear power plants.

Table F.6-1. State Environmental Requirements

Law/Regulation	Requirements
Air Quality Protection	
Title V Permit Rules	Establishes the policies and procedures by which a State will administer the Title V permit program under the CAA. Requires Title V sources to apply for and obtain a Title V permit prior to operation of the source facility.
Permits to Install New Sources of Pollution	Requires a permit prior to the installation of a new source of air pollutants or the modification of an air contaminant source. Discusses exemptions and conditions under which approval will be granted. Also requires an impact analysis to determine if the air contaminant source will cause or contribute to violations of the NAAQS.
Air Permits to Operate and Variances	Requires a permit prior to the operation or use of any air contaminant source in violation of any applicable air pollution control law, unless a variance has been applied for and obtained from the State agency.
Accidental Release Prevention Program	Requires the owner or operator of a stationary source, that has more than a threshold quantity of a regulated substance, to comply with all the provisions of the rule, including creating a hazard assessment, risk management plan, a prevention program, and an emergency response program.
General Conformity Rules	Rules on "general conformity" are mandated by the CAA to ensure that Federal actions do not contribute to air quality violations within the State. Discusses which Federal actions are subject to the conformity requirements, the procedures for conformity analysis, public participation/consultation, and the final conformity determination.
Water Resources Protection	
National Pollutant Discharge Elimination System Permits	Requires a permit prior to the discharge of pollutants from any point source into waters of the United States. Each permit holder must comply with authorized discharge levels, monitoring requirements, and other appropriate requirements in the permit.
Permits to Install New Sources of Pollution	Requires a permit prior to the installation of a new source of water pollutants or the modification of any pollutant discharge source.
Water Quality Standards	Establishes water quality standards for surface waters in the State, including beneficial use designations, numeric water quality criteria, and the anti-degradation water body classification system. Water quality standards are enforced through the NPDES permit.
Section 401 Water Quality Certifications	Requires a Section 401 water quality certification and payment of applicable fees before the issuance of any Federal permit or license to conduct any activity that may result in discharges to waters of the State.
Public Water Systems Licenses to Operate	Requires a public water system license prior to operating or maintaining a public water system.

Table F.6-1. (cont.)

Law/Regulation	Requirements
Water Resources Protection (cont.)	
Design, Construction, Installation, and Upgrading for Underground Storage Tank Systems	Establishes performance standards and upgrading requirements for underground storage tanks containing petroleum (e.g., diesel fuel) or other regulated substances. Requires an installation or upgrading permit for each location where such installation or upgrading is to occur prior to beginning either an installation or upgrading of a tank or piping comprising an underground storage tank system.
Registration of Underground Storage Tank System	Establishes annual registration requirements for underground storage tanks containing petroleum or other regulated substances.
Flammable and Combustible Liquids	Requires a permit to install, remove, repair, or alter a stationary tank for the storage of flammable or combustible liquids or modify or replace any line or dispensing device.
Waste Management and Pollution Prevention	
Generator Standards	Requires any person who generates waste to determine if that waste is hazardous. Requires a generator identification number from EPA or State agency prior to treatment, storage, disposal, transport, or offer for transport of hazardous waste.
Licensing Requirements for Solid Waste, Construction, and Demolition Debris Facilities	Requires an annual license for any municipal solid waste landfill, industrial solid waste landfill, residual solid waste landfill, compost facility, transfer facility, infectious waste treatment facility, or solid waste incineration facility prior to operation. New facilities must obtain a permit to install, prior to construction. Also, requires a license to establish, modify, operate, or maintain a construction and demolition debris facility.
Radiation Generator and Broker Reporting Requirements	Requires completion of a low-level radioactive waste generator report within 60 days of beginning to generate low-level waste. Additionally, requires each generator to submit an annual report on the state of low-level waste activities in their facility and pay applicable fees.
Hazardous Waste Management System Permits	Requires operation permits for any new or existing hazardous waste facility.
Emergency Planning and Response	
Hazardous Chemical Reporting	Requires the submission of Material Safety Data Sheets and an annual Emergency and Hazardous Chemical Inventory to local emergency response officials for any hazardous chemicals that are produced, used, or stored at the facility in an amount that equals or exceeds the threshold quantity.

Table F.6-1. (cont.)

Law/Regulation	Requirements
Emergency Planning and Response (cont.)	
Emergency Planning Requirements of Subject Facilities	Requires any facility having an extremely hazardous substance present in an amount equal to, or exceeding the threshold planning quantity, to notify the emergency response commission and the local emergency planning committee within 60 days after onsite storage begins. Also requires the designation of a facility representative who will participate in the local emergency planning process as a facility emergency coordinator.
Toxic Chemical Release Reporting	Establishes reporting requirements and schedule for each toxic chemical known to be manufactured (including imported), processed, or otherwise used in excess of an applicable threshold quantity. Applies only to facilities of a certain classification.
Biotic Resources Protection	
State Endangered Plant Species Protection	Establishes criteria for identifying threatened or endangered species of native plants and prohibits injuring or removing endangered species without permission.
State Endangered Fish and Wildlife Species Protection	Establishes and requires periodic update to a State list of endangered fish and wildlife species.
Permits for Impacts to Isolated Wetlands	Requires a general or individual isolated wetland permit prior to engaging in an activity that involves the filling of an isolated wetland.
Cultural Resources Protection	
State Registry of Archaeological Landmarks	Establishes a State registry of archaeological landmarks. Prohibits any person from excavating or destroying such land, or from removing skeletal remains or artifacts from any land, placed on the registry without first notifying the State Historic Preservation Office.
Survey and Salvage; Discoveries; Preservation	Directs State departments, agencies, and political subdivisions to cooperate in the preservation of archaeological and historic sites and the recovery of scientific information from such sites. Also, requires State agencies and contractors performing work on public improvements to cooperate with archaeological and historic survey and salvage efforts and to notify the State historic preservation office about archaeological discoveries.

F.6.2 Operating Permits and Other Requirements

Several operating permit applications may be prepared and submitted, and regulator approval and/or permits would be received, prior to license renewal approval by the NRC. Table F.6-2 lists representative Federal, State, and local permits.

Table F.6-2. Federal, State, and Local Permits and Other Requirements

License, Permit, or Other Required Approval	Responsible Agency	Authority	Relevance and Status
Air Quality Protection			
Title V Operating Permit: Required for sources that are not exempt and are major sources, affected sources subject to the Acid Rain Program, sources subject to new source performance standards, or sources subject to National Emission Standards for Hazardous Air Pollutants.	EPA or State agency	CAA, Title V, Sections 501–507 (USC, Title 42, Sections 7661-7661f [42 USC 7661-7661f])	Nuclear power plants are subject to 40 CFR Part 61, Subpart H (40 CFR Part 61, Subpart H), "National Emissions Standards for Emissions of Radionuclides," which is included in the terms and conditions of the Title V Operating Permit.
Risk Management Plan: Required for any stationary source that has a regulated substance (e.g., chlorine, hydrogen fluoride, nitric acid) in any process (including storage) in a quantity that is over the threshold level.	EPA or State agency	CAA, Title 1, Section 112(R)(7) (42 USC 7412)	These regulated substances stored in quantities that exceed the threshold levels would require a Risk Management Plan.
CAA Conformity Determination: Required for each criteria pollutant (i.e., sulfur dioxide, particulate matter, carbon monoxide, ozone, nitrogen dioxide, and lead) where the total of direct and indirect emissions in a nonattainment or maintenance area caused by a Federal action would equal or exceed threshold rates.	EPA or State agency	CAA, Title 1, Section 176(c) (42 USC 7506)	CAA conformity determination would be required at nuclear power plants located in nonattainment areas with NAAQS for criteria pollutants or maintenance areas for any criteria pollutant that would be emitted as a result of license renewal.
Water Resources Protection			
NPDES Permit: Construction Site Stormwater: Required before making point source discharges of stormwater from a construction project that disturbs more than 2 hectares (5 acres) of land.	EPA or State agency	CWA (33 USC 1251 et seq.); 40 CFR Part 122	Any plant refurbishment involving construction of more than 2 hectares (5 acres) of land would require a Stormwater Pollution Prevention Plan and construction site stormwater discharge permit.
NPDES Permit: Industrial Facility Stormwater: Required before making point source discharges of stormwater from an industrial site.	EPA or State agency	CWA (33 USC 1251 et seq.); 40 CFR Part 122	Stormwater would be discharged from the nuclear power plants during operations. Stormwater would discharge through existing outfalls covered by a permit.

Table F.6-2. (cont.)

License, Permit, or Other Required Approval	Responsible Agency	Authority	Relevance and Status
Water Resources Protection (cont.)			
NPDES Permit: Process Water Discharge: Required before making point source discharges of industrial process wastewater.	EPA or State agency	CWA (33 USC 1251 et seq.); 40 CFR Part 122	Process industrial wastewater would be discharged through existing outfalls covered by the permit.
Spill Prevention Control and Countermeasures Plan: Required for any facility that could discharge diesel fuel in harmful quantities into navigable waters or onto adjoining shorelines.	EPA or State agency	CWA (33 USC 1251 et seq.); 40 CFR Part 112	A Spill Prevention Control and Countermeasures Plan is required at nuclear power plants storing large volumes of diesel fuel and/or other petroleum products.
CWA Section 401 Water Quality Certification: Required to be submitted to the agency responsible for issuing any Federal license or permit to conduct an activity that may result in a discharge of pollutants into waters of a State.	EPA or State agency	CWA, Section 401 (33 USC 1341); ORC Chapters 119 and 6111	Certification for operation of a nuclear power plant may require a Federal license or permit (e.g., a CWA Section 404 Permit).
New Underground Storage Tanks System Registration: Required within 30 days of bringing a new underground storage tank system into service.	EPA or State agency	RCRA, as amended, Subtitle I (42 USC 6991a–6991i); 40 CFR 280.22	Required if new underground storage tank systems would be installed at a nuclear power plant.
Above Ground Storage Tank: A permit is required to install, remove, repair, or alter any stationary tank for the storage of flammable or combustible liquids.	State Fire Marshal		Required if new aboveground diesel fuel storage tanks would be installed at a nuclear power plant.
Waste Management and Pollution Prevention			
Registration and Hazardous Waste Generator Identification Number: Required before a person who generates over 100 kg (220 lb) per calendar month of hazardous waste ships the hazardous waste offsite.	EPA or State agency	RCRA, as amended (42 USC 6901 et seq.), Subtitle C	Generators of hazardous waste must notify the EPA that the wastes exist and require management in compliance with RCRA.

Table F.6-2. (cont.)

License, Permit, or Other Required Approval	Responsible Agency	Authority	Relevance and Status
Waste Management and Pollution Prevention (cont.)			
Hazardous Waste Facility Permit: Required if hazardous waste will undergo nonexempt treatment by the generator, be stored onsite for longer than 90 days by the generator of 1,000 kg (2,205 lb) or more of hazardous waste per month, be stored onsite for longer than 180 days by the generator of between 100 and 1,000 kg (220 and 2,205 lb) of hazardous waste per month, disposed of onsite, or be received from offsite for treatment or disposal.	EPA or State agency	RCRA, as amended (42 USC 6901 et seq.), Subtitle C	Hazardous wastes are usually not disposed of onsite at nuclear power plants. Hazardous wastes generated onsite are not generally stored for more than 90 days. However, should a nuclear power plant store waste onsite for greater than 90 days for characterization, profiling, or scheduling for treatment or disposal, a Hazardous Waste Facility Permit would be required.
Emergency Planning and Response			
List of Material Safety Data Sheets: Submission of a list of Material Safety Data Sheets is required for hazardous chemicals (as defined in 29 CFR Part 1910) that are stored onsite in excess of their threshold quantities.	State and local emergency planning agencies	EPCRA, Section 311 (42 USC 11021); 40 CFR 370.20	Nuclear power plant operators are required to submit a list of Material Safety Data Sheets to State and local emergency planning agencies.
Annual Hazardous Chemical Inventory Report: The report must be submitted when hazardous chemicals have been stored at a facility during the preceding year in amounts that exceed threshold quantities.	State and local emergency response agencies; local fire department	EPCRA, Section 312 (42 USC 11022); 40 CFR 370.25	If hazardous chemicals have been stored at a nuclear power plant during the preceding year in amounts that exceed threshold quantities, then plant operators would be required to submit an annual Hazardous Chemical Inventory Report.
Notification of Onsite Storage of an Extremely Hazardous Substance: Submission of the notification is required within 60 days after onsite storage begins of an extremely hazardous substance in a quantity greater than the threshold planning quantity.	State and local emergency response agencies	EPCRA, Section 304 (42 USC 11004); 40 CFR 355.30	If an extremely hazardous substance will be stored at a nuclear power plant in a quantity greater than the threshold planning quantity, plant operators would prepare and submit the Notification of Onsite Storage of an Extremely Hazardous Substance.

Table F.6-2. (cont.)

License, Permit, or Other Required Approval	Responsible Agency	Authority	Relevance and Status
Emergency Planning and Response (cont.)			
Annual Toxics Release Inventory Report: Required for facilities that have 10 or more full-time employees and are assigned certain Standard Industrial Classification Codes.	EPA or State agency	EPCRA, Section 313 (42 USC 11023); 40 CFR Part 372	If required, nuclear power plant operators would prepare and submit a Toxics Release Inventory Report to the EPA.
Transportation of Radioactive Wastes and Conversion Products Packaging, Labeling, and Routing Requirements for Radioactive Materials: Required for packages containing radioactive materials that will be shipped by truck or rail.	USDOT	Hazardous Materials Transportation Act (49 USC 1501 et seq.); AEA, as amended (42 USC 2011 et seq.); 49 CFR Parts 172, 173, 174, 177, and 397	When shipments of radioactive materials are made, nuclear power plant operators would comply with USDOT packaging, labeling, and routing requirements.
Biotic Resource Protection			
Threatened and Endangered Species Consultation: Required between the responsible Federal agencies and USFWS and/or NMFS to ensure that the project is not likely to: (1) jeopardize the continued existence of any species listed at the Federal or State level as endangered or threatened, or (2) result in destruction of critical habitat of such species.	USFWS and NMFS	ESA of 1973, as amended (16 USC 1531 et seq.)	For actions that may affect listed species or designated critical habitat, the NRC would consult with the USFWS and/or NMFS under Section 7 of the ESA.
Essential Fish Habitat Consultation: Required between the responsible Federal agency and NMFS to ensure that Federal actions authorized, funded, or undertaken do not adversely affect essential fish habitat.	NMFS	MSA, as amended (16 USC 1801-1884)	For actions that may adversely affect essential fish habitat, the NRC would consult with NMFS in accordance with 50 CFR Part 600, Subpart J.

Table F.6-2. (cont.)

License, Permit, or Other Required Approval	Responsible Agency	Authority	Relevance and Status
Biotic Resource Protection (cont.)			
CWA Section 404 (Dredge and Fill) Permit: Required to place dredged or fill material into waters of the United States, including areas designated as wetlands, unless such placement is exempt or authorized by a nationwide permit or a regional permit; a notice must be filed if a nationwide or regional permit applies.	USACE	CWA (33 USC 1251 et seq.); 33 CFR Parts 323 and 330	Any dredging or placement of fill material into wetlands within the jurisdiction of the USACE at a nuclear power plant would require a Section 404 permit.
Cultural Resources Protection			
Archaeological and Historical Resources Consultation: Required before a Federal agency approves a project in an area where archaeological or historic resources might be located.	State Historic Preservation Officer and/or Tribal Historic Preservation Officer	NHPA of 1966, as amended (16 USC 470 et seq.); Archaeological and Historical Preservation Act of 1974 (16 USC 469-469c-2); Antiquities Act of 1906 (16 USC 431 et seq.); Archaeological Resources Protection Act of 1979, as amended (16 USC 470aa–mm)	The NRC would consult with the State and/or Tribal Historic Preservation Officers and representative Indian Tribes regarding the impacts of license renewal and the results of archaeological and architectural surveys of nuclear power plant sites.

F.7 Emergency Management and Response Laws, Regulations, and Executive Orders

This section discusses the response laws, regulations, and Executive Orders that address the protection of public health and worker safety and require the establishment of emergency plans. These laws, regulations, and Executive Orders relate to the operation of nuclear power plants. For ease of the reader, certain items are repeated from previous sections in this appendix.

F.7.1 Federal Emergency Management Response Laws

Emergency Planning and Community Right-to-Know Act of 1986 (42 USC 11001 et seq.) (also known as "SARA Title III")—The Emergency Planning and Community Right-to-Know

Act of 1986 (EPCRA), which is the major amendment to CERCLA (42 USC 9601*)*, establishes the requirements for Federal, State, and local governments, Indian Tribes, and industry regarding emergency planning and "Community Right-to-Know" reporting on hazardous and toxic chemicals. The "Community Right-to-Know" provisions increase the public's knowledge and access to information on chemicals at individual facilities, their uses, and releases into the environment. States and communities working with facilities can use the information to improve chemical safety and protect public health and the environment. This Act requires emergency planning and notice to communities and government agencies concerning the presence and release of specific chemicals. The EPA implements this Act under regulations found in 40 CFR Parts 355, 370, and 372.

Comprehensive Environmental Response, Compensation, and Liability Act of 1980 (42 USC 9604(l) (also known as "Superfund")—This Act provides authority for Federal and State governments to respond directly to hazardous substance incidents. The Act requires reporting of spills, including radioactive spills, to the National Response Center.

Robert T. Stafford Disaster Relief and Emergency Assistance Act of 1988 (42 USC 5121)—This Act, as amended, provides an orderly, continuing means of providing Federal Government assistance to State and local governments in managing their responsibilities to alleviate suffering and damage resulting from disasters. The President, in response to a State governor's request, may declare an "emergency" or "major disaster" to provide Federal assistance under this Act. The President, in Executive Order 12148, delegated all functions except those in Sections 301, 401, and 409 to the Director of the Federal Emergency Management Agency. The Act provides for the appointment of a Federal coordinating officer who will operate in the designated area with a State coordinating officer for the purpose of coordinating State and local disaster assistance efforts with those of the Federal Government.

Justice Assistance Act of 1984 (42 USC 3701–3799)—This Act establishes emergency Federal law enforcement assistance to State and local governments in responding to a law enforcement emergency. The Act defines the term "law enforcement emergency" as an uncommon situation which requires law enforcement, which is or threatens to become of serious or epidemic proportions, and with respect to which State and local resources are inadequate to protect the lives and property of citizens or to enforce the criminal law. Emergencies that are not of an ongoing or chronic nature (for example, the Mount St. Helens volcanic eruption) are eligible for Federal law enforcement assistance including funds, equipment, training, intelligence information, and personnel.

Price-Anderson Nuclear Industries Indemnity Act (42 USC 2210)—The Price-Anderson Act provides insurance protection to victims of a nuclear accident. The main purpose of the Act is to partially indemnify the nuclear industry against liability claims arising from nuclear incidents while still ensuring compensation coverage for the general public. The Act establishes a no-fault insurance-type system in which the first $12.6 billion (as of 2011) is industry-funded as described in the Act (any claims above the $12.6 billion would be covered by the Federal Government).

The Act requires NRC licensees and U.S. Department of Energy contractors to enter into agreements of indemnification to cover personal injury and property damage to those harmed by a nuclear or radiological incident, including the costs of incident response or precautionary evacuation, costs of investigating and defending claims, and settling suits for such damages.

F.7.2 Federal Emergency Management and Response Regulations

Quantities of Radioactive Materials Requiring Consideration of the Need for an Emergency Plan for Responding to a Release (10 CFR 30.72, Schedule C)—This section of the regulations provides a list that is the basis for both the public and private sector to determine whether the radiological materials they handle must have an emergency response plan for unscheduled releases. The "Federal Radiological Emergency Response Plan," dated November 1995, primarily discusses offsite Federal response in support of State and local governments with jurisdiction during a peacetime radiological emergency.

Occupational Safety and Health Administration Emergency Response, Hazardous Waste Operations, and Worker Right to Know (29 CFR Part 1910)—This regulation establishes OSHA requirements for employee safety in a variety of working environments. It addresses employee emergency and fire prevention plans (Section 1910.38), hazardous waste operations and emergency response (Section 1920.120), and hazards communication (Section 1910.1200) to make employees aware of the dangers they face from hazardous materials in their workplace. These regulations do not directly apply to Federal agencies. However, Section 19 of the Occupational Safety and Health Act (29 USC 668) requires all Federal agencies to have occupational safety programs "consistent" with Occupational Safety and Health Act standards.

Emergency Management and Assistance (44 CFR Section 1.1)—This regulation contains the policies and procedures for the Federal Emergency Management Act, National Flood Insurance Program, Federal Crime Insurance Program, Fire Prevention and Control Program, Disaster Assistance Program, and Preparedness Program, including radiological planning and preparedness.

Hazardous Materials Tables and Communications, Emergency Response Information Requirements (49 CFR Part 172)—This regulation defines the regulatory requirements for marking, labeling, placarding, and documenting hazardous material shipments. The regulation also specifies the requirements for providing hazardous material information and training.

F.7.3 Emergency Management and Response Executive Orders

Executive Order 12148, Federal Emergency Management—This Order transfers functions and responsibilities associated with Federal emergency management to the Director of the Federal Emergency Management Agency. The Order assigns the Director the responsibility to establish Federal policies and to coordinate all civil defense and civil emergency planning for the management, mitigation, and assistance functions of Executive agencies.

Executive Order 12656, Assignment of Emergency Preparedness Responsibilities—This Order assigns emergency preparedness responsibilities to Federal departments and agencies.

Executive Order 12938, Proliferation of Weapons of Mass Destruction—This Order states that the proliferation of nuclear, biological, and chemical weapons ("weapons of mass destruction") and the means of delivering such weapons constitutes an unusual and extraordinary threat to the national security, foreign policy, and economy of the United States, and that a national emergency would be declared to deal with that threat.

F.8 Consultations with Agencies and Federally Recognized American Indian Nations

Certain laws, such as the ESA, the Fish and Wildlife Coordination Act, and the NHPA, require consultation and coordination by the NRC with other governmental entities including other Federal, State, and local agencies and Federally recognized American Indian governments. These consultations must occur on a timely basis and are generally required before any land disturbance can begin. Most of these consultations are related to biotic resources, cultural resources, and American Indian rights. The biotic resource consultations generally pertain to the potential for activities to disturb sensitive species or habitats. Cultural resource consultations relate to the potential for disruption of important cultural resources and archaeological sites. American Indian consultations are concerned with the potential for disturbance of ancestral American Indian sites, the traditional practices of American Indians, and natural resources of importance to American Indians.

NRC FORM 335
(12-2010)
NRCMD 3.7

U.S. NUCLEAR REGULATORY COMMISSION

BIBLIOGRAPHIC DATA SHEET

(See instructions on the reverse)

1. REPORT NUMBER
(Assigned by NRC, Add Vol., Supp., Rev., and Addendum Numbers, if any.)
NUREG-1437, Volume 3, Revision 1

2. TITLE AND SUBTITLE
Generic Environmental Impact Statement for License Renewal of Nuclear Plants
Appendices
Final Report

3. DATE REPORT PUBLISHED

MONTH	YEAR
June	2013

4. FIN OR GRANT NUMBER

5. AUTHOR(S)
See Chapter 5 of NUREG-1437, Volume 1

6. TYPE OF REPORT
Technical

7. PERIOD COVERED (Inclusive Dates)

8. PERFORMING ORGANIZATION - NAME AND ADDRESS (If NRC, provide Division, Office or Region, U. S. Nuclear Regulatory Commission, and mailing address; if contractor, provide name and mailing address.)
Division of License Renewal
Office of Nuclear Reactor Regulation
U.S. Nuclear Regulatory Commission
Washington, D.C. 20555-0001

9. SPONSORING ORGANIZATION - NAME AND ADDRESS (If NRC, type "Same as above", if contractor, provide NRC Division, Office or Region, U. S. Nuclear Regulatory Commission, and mailing address.)
Same as 8 above

10. SUPPLEMENTARY NOTES

11. ABSTRACT (200 words or less)
U.S. Nuclear Regulatory Commission (NRC) regulations allow for the renewal of commercial nuclear power plant operating licenses. To support the license renewal environmental review process, the NRC published the Generic Environmental Impact Statement for License Renewal of Nuclear Plants (GEIS) in 1996. The proposed action considered in the GEIS is the renewal of nuclear power plant operating licenses.
Since publication of the GEIS, approximately 40 plant sites (70 reactor units) have applied for license renewal and undergone environmental reviews, the results of which were published as supplements to the 1996 GEIS. This GEIS revision reviews and reevaluates the issues and findings of the 1996 GEIS. Lessons learned and knowledge gained during previous license renewal reviews provide a significant source of new information for this assessment. In addition, new research, findings, public comments, and other information were considered in evaluating the significance of impacts associated with license renewal.
The intent of the GEIS is to determine which issues would result in the same impact at all nuclear power plants and which issues could result in different levels of impact at different plants and thus require a plant-specific analysis for impact determinations. The GEIS revision identifies 78 environmental impact issues for consideration in license renewal environmental reviews, 59 of which have been determined to be generic to all plant sites. The GEIS also evaluates a full range of alternatives to the proposed action. For most impact areas, the proposed action would have impacts that would be similar to or less than impacts of the alternatives, in large part because most alternatives would require new power plant construction, whereas the proposed action would not.

12. KEY WORDS/DESCRIPTORS (List words or phrases that will assist researchers in locating the report.)
Generic Environmental Impact Statement for License Renewal of Nuclear Plants
GEIS
Generic-1437, Revision 1
National Environmental Policy Act
NEPA
License Renewal

13. AVAILABILITY STATEMENT
unlimited

14. SECURITY CLASSIFICATION
(This Page)
unclassified
(This Report)
unclassified

15. NUMBER OF PAGES

16. PRICE

UNITED STATES
NUCLEAR REGULATORY COMMISSION
WASHINGTON, DC 20555-0001

OFFICIAL BUSINESS

NUREG-1437, Vol. 3
Revision 1

Generic Environmental Impact Statement for License Renewal of Nuclear Plants

June 2013

www.ingramcontent.com/pod-product-compliance
Lightning Source LLC
Chambersburg PA
CBHW080233180526
45167CB00006B/2259
* 9 7 8 1 4 9 9 6 3 6 5 6 7 *